CONTRIBUTIONS TO HUMAN BIOMETEOROLOGY

Edited by W. Selvamurthy

PROGRESS IN BIOMETEOROLOGY
EDITOR H. LIETH

VOLUME 4

SPB Academic Publishing 1987

ISBN 90-5103-004-5

CIP-DATA KONINKLIJKE BIBLIOTHEEK, DEN HAAG

Contributions

Contributions to human biometeorology / ed. by
W. Selvamurthy. – Den Haag: SPB Academic Publishing. –
(Progress in biometeorology; vol. 4)
With index, ref.
ISBN 90-5103-004-5 bound
SISO 556 UDC 551.586:612 NUGI 819
Subject heading: physiology; biometeorology.

CONTENTS

Page

EDITORIAL NOTE

The knowledge of the impact of climate and weather on human health and efficiency has recently assumed immense significance in planning growth and development, particularly in Developing countries. Considering the global emphasis on this problem, the focal theme of the International Conference on Biometeorology held at New Delhi during 1983 was rightly chosen to be 'Biometeorology for Development'.

Several important research contributions on human biometeorology were presented in the symposium on 'Climate, Human Health and Efficiency', and related symposia. Some of the papers have been selected and updated for publication in this volume.

The contributions cover wide ranging topics which include physiology of adaptation to environments such as high altitude, desert, cold, and electromagnetic fields, and biometeorology of urbanization and housing. This compilation of recent work on human biometeorology will be very useful to biologists, meteorologists, urban and town planners, physicians and psychologists, who are interested in understanding the interaction of the environment with human health and efficiency.

W. Selvamurthy

New Delhi

INFLUENCE OF URBANIZATION ON LOCAL TEMPERATURE AND HUMIDITY FIELDS IN A FEW MAJOR CITIES IN INDIA

A.K. Mukherjee, B. Mukhopadhyay
and Krishna Nand
(Meteorological Office, Pune)

Abstract: - The influences of geographic location and urban configuration on the characteristics of heat-islands in some Indian cities have been discussed. The conclusions have been utilised to comment on aspects of urban planning.

INTRODUCTION

Urbanization modifies the climate of a city to a large extent. Of all the meteorological parameters that are affected by urbanization, temperature anomalies have been studied in great detail by various workers (Chandler, 1965; Nkemdirm, 1976; Oke and Maxwell, 1975; Mukherjee and Daniel, 1976; Padmanabhamurty and Bahl, 1979; Krishna Nand and Maske, 1981). The positive temperature anomaly of the built-up area with respect to the rural, called the urban heat-island, has a very limited vertical extent and is most intense during calm winter nights. A detailed study of this phenomenon for different cities in India will help in better urban planning as well as in assessing the likely dispersal of pollutants within the city. The cities included in this study are Bombay, Calcutta, Delhi and Pune, all of which are large and densely populated.

METHODS

Temperature/humidity fields in different cities were obtained by mobile surveys within the city; Charts were then analysed for a fixed instant of time. Daily minimum and maximum temperature data were also obtained from a network of 11 stations in Delhi.

RESULTS AND DISCUSSIONS

Studies at Bombay: The city of Bombay is a north-south oriented peninsula jutting into the Arabian Sea with its northern end in continuity with the mainland of the Indian west coast. Studies conducted by Daniel and Krishna Murty (1973), Philip et al. (1973) and Mukherjee and Daniel (1976) reveal that the heat-island is formed near the southern tip, where the density of concrete structures is maximum. The coldest pocket is in the north near the foothills of Jogeshwari (J). The population density is lowest at this place and a katabatic effect further feeds these places with cold air. The most intense heat island of intensity 11° C on the night of 11th January 1975 was reported by Mukherjee et al. (1976) (Fig. 1). The heat island stretched from the warmest pocket in Colaba (C) to Dadar (D). A cold tongue extended from Jogeshwari to regions as far south as Grant Road (G). This indicates that the cold air has a tendency to move towards the warm regions thereby bringing suburban air to the city.

Seasonal variation: The data indicate that the locations of heat-islands do not appreciably change with seasons. Intensities are maximum in winter (5° C) and minimum in Monsoon (2° C). The reasons for low values during the latter season are increased cloud cover, moisture content and windiness.

Maritime influence: The city is influenced by the warmer sea in winter from either side. An east-west section through the temperature field shows a pronounced peak at the centre (heat-island) and two secondary ones on either side (Fig. 1).

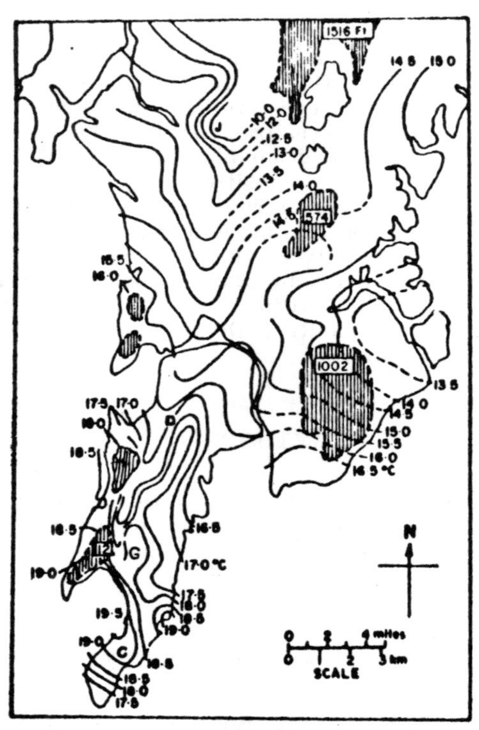

Figure 1: *Isotherms at Bombay, 11th January 1975 0600 I.S.T. (After: Mukherjee et al. 1976).*

Influence of wind: It is known from the results of Chandler (1965) and Bornstein et al. (1972) that for very low rural wind speed conditions during the night the urban flow is stronger. It is caused by an intense heat-island circulation while for weak heat-island conditions the situation is the reverse. A critical speed of about 15 km/h results in equal windiness in the surrounding of the city area. It may thus be stated that at Bombay, when the easterly land breeze is aided by an easterly flow on the synoptic scale, the upwind rural speeds may exceed the critical speed and erode the heat island considerably. A situation contrary to this would help in intensifying the heat island. The synoptic situation on the night of 11th January 1975 caused a westerly flow at lower levels which did not allow the land breeze to set in. This resulted in an exceptionally intense heat island on that night. It may be pointed out here that during winter nights the rural air flow in Bombay has a major easterly component as the tallest structures are aligned along the west coast near Colaba. The obstructed drainage due to this causes retention of warmth on the windward side.

Studies at Calcutta: Calcutta is situated on the eastern bank of Hoogly river and is 80 km inland from the Bay of Bengal. Population density is very high in the northern regions. The southwestern localities near the docks are also densely populated. The central part of the city has a shallow basin like topography.

Two surveys were conducted on the nights of 28th February 1978 and 15th January 1983 respectively. On the earlier occasion two heat islands were noticed, one in the north, along the bank of the river (5° C) and the other, an extended one, in the southern region (4° C). During the second survey only the southern one was noticed (Fig. 2) which itself was restricted to the south-west sector. The surface wind during early morning of the earlier survey was NNW, 4 km/h which advected some warmth from the river and caused the heat-island along the northern bank of the river. On the second occasion the wind was NE, 9 km/h, which was dry and had a ventilating effect. Hence the northern warm pocket was eroded. In Fig. 2 it can be noticed that the dew points near the cold pocket were less than those near the heat island indicating that the cold dry air which blew from NE was unable to penetrate deep into the south-western regions where the urban canopy layer retained its humidity mixing ratio. In regions where the dry and cold air penetrated quite deep due to the general slope of terrain, the dew points were found much less (e.g. central Calcutta). The deep cleft in isotherms on the east-central regions also indicate an inflow of cold suburban air from the east into the basin like topography of Central Calcutta.

Studies at Delhi: Delhi is a sprawling metropolis in the northwest interior of the subcontinent, having a subtropical type of winter, extending from December to March during which low level inversions are frequent.

Surveys carried out by Bahl and Padmanabhamurty (1979) and Padmanabhamurty and Bahl (1980) reveal that major zones of warmth develop in the northern parts of Delhi near the Subzi Mandi area (Fig. 3), either as a single heat-island or as multiple ones, each roughly of

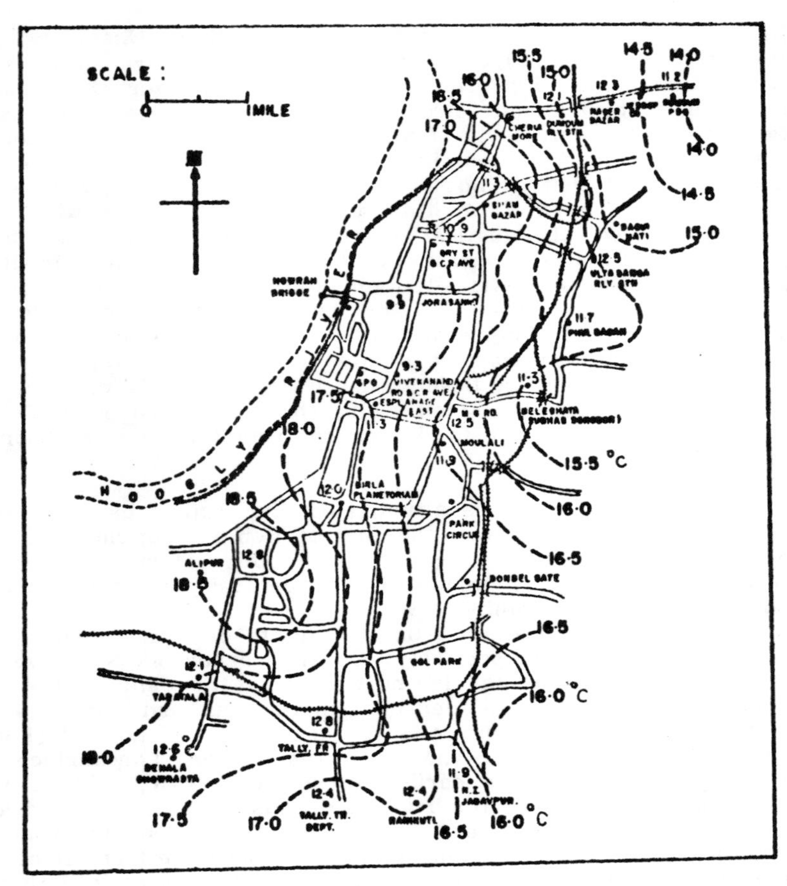

Figure 2: *Isotherms and spot values of dew point temperature at Calcutta 15th January 1983 min. temperature epoch.*

intensity of 5° C to 6° C. One of these heat-islands forms close to Jamuna river, with its axis oriented along the river. This is indicative of the influence of the relative warmth of the river. In the south and west there are other low intensity heat islands. Running east-west through the central regions of the city lays a cold tongue on all occasions. It coincided with the location of a hilly ridge which is completely uninhabited. In general, it can be remarked that in a city having a radial distribution of places of high population density interspersed with open regions, no single large heat island would form but several pockets of warmth and coolness may be observed.

Climatological features: Daily maximum and minimum data were obtained for the whole year from a number of Urban Climatological stations within the city. Minimum temperature data were utilized by

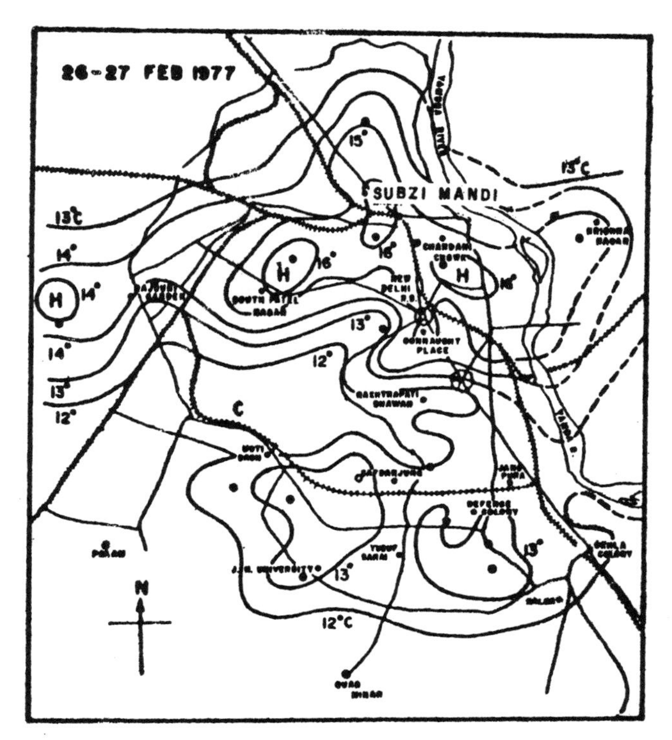

Figure 3: *Isotherms at Delhi, 27th February 1977 minimum temperature epoch, clear sky and calm wind condition. (After: Bahl and Padmanabhamurty, 1979).*

Krishna Nand and Maske (1981) to delineate the heat islands. In the monthly means only the heat island over North-Delhi (maximum intensity 5.1° C in March and a minimum of 0.8° C in July) was found and the large cold pocket over the usual place on the ridge was also seen. Other small features were not reflected in these due to the variations in their day to day locations. The climatologically evaluated temperature anomaly for the day time using maximum temperature was about 1° C. This illustrates the fact that mixing of the surface air ruptures the conservative urban canopy layer during the day and homogenises the rural and urban atmospheres.

Studies at Pune: The city of Pune is on the Deccan Plateau and is skirted by hills on the western and south-western sides. The slope of the terrain is generally from south-west to north-east.

Influence of expansion of human settlements: In the earliest survey of 1972, Daniel and Krishnamurty (1973) reported the formation of only one largest heat island in the southern part of the city (intensity 4 to 6° C during different nights in Feb. 1972), which has the highest density of population and urban dwellings. All other studies including a survey by the present authors on 23rd and 24th January 1982 indicate many other heat islands. These heat islands have shown a tendency over the years, to migrate slightly northwards where population density and number of concrete structures have been increasing.

Terrain effect: The down slope movement of cold air is most pronounced for Pune city (Fig. 4) as compared to those discussed earlier. The penetration of colder air from southwest along the river bed causes the splitting of the large heat island in the south into two separate islands (Phule Market and Deccan Gymkhana). The cold pocket at the Central Observatory, situated in a vast vegetated expanse, did not penetrate southwards because of the opposing gradient of terrain.

Effect of river: Only small zones of warmth were observed near the river. The river is able to influence the downwind localities by an increase in humidity mixing ratios. It was found that moisture advection from the river is normally negligible in the inner parts of the city.

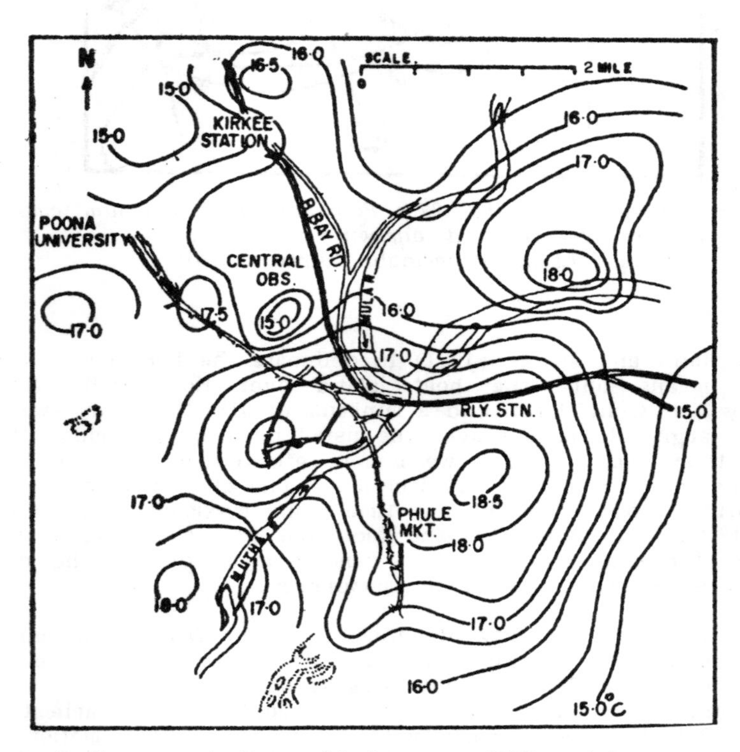

Figure 4: *Isotherms at Pune 24 January 1982, minimum temperature epoch.*

COMPARATIVE STUDY OF THE DATA FROM ALL CITIES

A few salient features are highlighted here regarding modification in the temperature field in different cities in India due to urbanisation.
(a) In a maritime city like Bombay, lower intensity heat islands are expected than continental ones, but inspite of that one does observe the formation of some, very intense heat islands, under favourable synoptic meteorological conditions which prevent the setting up of the ventilating land breeze.
(b) Radially spread out cities, like Delhi, have multiple heat islands, none of which are large in extent and they may inhibit the generation of intense circulation associated with heat islands.
(c) Low gradients in dew point fields normally seen over inland cities suggest that heat island induced flow can have little effect on Relative Humidity patterns unless affected by an upstream water body.

FEEDBACK INTO URBAN PLANNING

(a) For a city with a large population and massive built-up area, attempts must be made to break them in to smaller units with interspersed open spaces, so that one single heat island may not form and concentration of pollutants at one place may not occur.
(b) For coastal cities the land breeze has a ventilating effect. Thus, tall structures obstructing the drainage should not align the coast (transverse to the wind).

Over and above these, consideration of human comfort should also be incorporated in the basic planning of urban localities. For a place with dry winter (Delhi), discomfort increases due to low Relative Humidity. This can be compensated by providing large parks and water bodies in between pockets of dense settlements. This would also help in making summer nights more comfortable. The effect of increased humidity may, however, shift the comfort index adversely during summer days, thereby calling for a need to optimise this factor. For a place with a moist climate the only alternative is to allow proper ventilation by suburban air.

CONCLUSION

It has been found that heat island intensities are between 4 and 6° C on individual nights, as well as in monthly means, but heat islands as intense as even 12° C have been observed on individual occasions. The intensities depend on the urban characteristics, atmospheric stability conditions, wind speeds and also presence of water bodies. Intense heat islands can develop circulations which are capable of mass transport in the horizontal plane. Hence study of these are important from the point of view of pollutant dispersal. Even relative humidity fields are affected due to urbanization and a lowering by about 30% is noticeable over the warm pockets. Study of the above data suggests that modification in temperature and humidity fields are related to population and building density, topography and vegetation. These results can be used for better town planning.

ACKNOWLEDGEMENTS

The authors are very much thankful to staff members of the Meteorological Office in Pune for their assistance in conducting the mobile surveys at Pune.

REFERENCES

BAHL, H.D. and PADMANABHAMURTY, B. (1979): Heat island Studies at Delhi. Mausam, 30: 119-122.

BORNSTEIN, R.D., LORENZ, A. and JOHNSON, D.S. (1972): Recent Observations on urban effects on winds and temperatures in and around New York. Preprint, Conference on Urban Environment and Second Conference on Biometeorology; American Met. Soc. Philadelphia, pp. 89-94.

CHANDLER, T.J. (1965): The Climate of London. Hutchinson London, pp. 292. Reviewed in WMO technical Note 149.

DANIEL, C.R.J. and KRISHNAMURTY, K. (1973): Urban Temperature fields at Pune, and Bombay. Indian. J. Met. Geophys. 24: 407-412.

KRISHNA NAND and MASKE, S.J. (1981): Mean Heat island intensities at Delhi assessed from Urban Climatological data. Mausam 32: 269-272.

MASKE, S.J., KRISHNA NAND, BEHERE, P.G. and KACHARE, S.D. (1978): Characteristics of Heat Island at Pune. Prepubl. Sc. Report, India Meteorological Department, 78/12.

MUKHERJEE, A.K. and DANIEL, G.E.J. (1976): Temperature distribution of Bombay during a cold night. Indian J. Met. Geophysics, 27: 37-41.

NKEMDIRM, L.C. (1976): Dynamics of an Urban temperature field - A case study. J. Appl. Meteorol. 15: 818-828.

OKE, T.R. and MAXWELL, G.B. (1975): Urban Heat-Island dynamics in Montreal and Vancouver. Atmos. Environ. 9: 191-200.

PADMANABHAMURTY, B. (1979): Isotherms and isohumes in Pune on clear winter nights: A meso meteorological study. Mausam 30: 134-138.

PADMANABHAMURTY, B. and BAHL, H.D (1980): On surface structure and movement of heat islands and humidity islands at Delhi. Proceedings of symp. on Management of Environment. (Ed.) B. Patel, February 1980, Bombay.

PADMANABHAMURTY, B. and BAHL, H.D. (1981): Eco-climatic modification of Delhi due to urbanization. Mausam, 32: 205-300.

PHILIP, N.M. and KRISHNAMURTY, K. (1973): Seasonal Variation of Surface Temperature distribution over Bombay. Proceedings of Symposium on Environmental Pollution, Nagpur.

DUST STORMS AND ASSOCIATED WEATHER CHANGES IN AN ARID ENVIRONMENT

Y.S. Ramakrishna, G.G.S.N. Rao and B.V. Ramana Rao
(Central Arid Zone Research Institute, Jodhpur)

Abstract: - The vulnerability of the arid regions of N.W. India to dust storm and wind erosion activity have been discussed and the frequency of occurrence of dust storms over N.W. India during 1955-1965 were compared to those during 1931-1940. The influence of dust storm activity on the radiation characteristics of the atmosphere and also on the thermal regime and evaporative demand of the atmosphere were analysed. Methods suitable for minimising dust storm activity have been suggested.

INTRODUCTION

The arid regions of Northwest India are plagued by the age-old problem of dust storm activity during the summer and pre-monsoon seasons which often lead to wind erosion on a large scale. Dry and structureless top soil, lack of protective vegetative cover during the summer months experiencing high winds speeds, extended dry periods and over-grazing of land also contribute to this problem by providing loose and dry soil particles of the size that can be lifted by air, leading to increased dust storm activity. Paleobotanical evidence (Vishnu Mittre, 1977) suggests that man in the Rajasthan desert had been forced to live with the dust and dust storm activity over the region, from a long time ago. Measurements of dust concentration during summer over the Thar desert region (Bryson, 1971) indicate that average dust concentrations of the order of 300 to 800 $\mu g/m^3$ in the lowest 5 to 10 km of air. Compared to the turbid air over Chicago (150-200 $\mu g/m^3$), the air over the Thar desert especially during the summer months can be considered as very turbid.

The dust storm activity as a major weather hazard not only disturbs the ecological balance of the arid region of N.W. India but also creates weather changes in the arid environment, as the large quantities of dust suspended in the air for considerable time, influence the radiation characteristics and energy balance parameters of the atmosphere. The thermal and the moisture regime of the atmosphere also are considerably influenced by dust storm activity, though for a short period of time.

FREQUENCY OF DUST STORM ACTIVITY OVER N. INDIA

Frequencies of dust storm activity per year at 67 stations over N.India for the period 1955-1965 were collected from I.M.D. records and analysed. The data interestingly revealed that the frequency of dust storm activity is lowest in the Jaisalmer region which lies in the extreme arid region of the Indian desert and also over the Gujarat region, inspite of the Wind regime during the summer months in these regions remaining quite strong. The cores of maxima in dust storm frequency are observed over the region around Jaipur and the other over North Western region covering between Delhi and Amritsar, the core being at its maximum (13 per year) at Amritsar. Major part of N.India on the average records around 5 dust storms per year while the northwestern region records about 10 dust storms per year.

To analyse whether the dust storm activity has increased or decreased during recent times, the frequency of dust storm activity over N.India during the 10 year period 1955-1965 was compared with the dust storm activity during 1931 to 1940. The data clearly revealed that the dust storm activity during the 1930's was indeed double to that occurred during the period 1955-1965. Also the core of maximum dust storm activity during 1931-1940 was over Bikaner and Ganganagar region but a northward shift in the core over to Ganganagar, Amritsar region was discernable during 1955-1965. Even though the data used for comparision is less, it shows that dust storm activity has slightly decreased over N.W.India which can be attributed to more stabilization of sand dues and increased afforestation and grassland management systems adopted over the region.

DUST STORM ACTIVITY AND RAINFALL

Studies carried out on the monsoon rainfall and the frequency of dust storms during the subsequent year in the Indian arid zone indicated that dust storms followed a pattern similar to that of monsoon activity (Mann, 1979; Mann and Ramakrishna, 1980). Whenever the rainfall during the monsoon season was very low, there was a sharp increase in dust storm activity in the subsequent year and vice versa. Goudie (1978) worked out the relationship between mean annual precipitation and mean annual number of dust storms over India based on 101 observations and indicated a linear correlation of 0.42 significant at 1% level between the two. Ramakrishna et al. (1983) showed that the ratio of $\frac{E/P}{E/P(\text{normal})}$, where E is the open pan evaporation and P is the monthly precipitation respectively during the months of July and August, gives a satisfactory correlation (0.76) with the number of dust storms during the subsequent year.

INFLUENCE ON RADIATIVE CHARACTERISTICS

Atmospheric dust can influence the radiation budget of the atmosphere in two ways. It can directly influence by changing the radiation fluxes in a cloud-free atmosphere by scattering and absorption of radiation and indirectly by modifying the optical properties, particularly the albedo of clouds (Junge, 1979). The total incoming solar radiation on a dusty day and a clear (non-dusty) day at Jodhpur are presented in Fig. 1. It can be clearly seen that there is a strong depletion of the incoming radiation due to absorption and scattering in the atmosphere caused by the presence of dust. There was a decrease of 48% of the total incoming radiation on a dusty day, the greatest depletion during short spells reaching as high as 80%.

INFLUENCE ON THERMAL REGIME

Large quantities of dust suspended in the air not only reduce the incoming solar radiation but also influence the outgoing longwave radiation from the ground during the night period. The result would be a decreased diurnal temperature range on a dusty day. Data on the maximum and minimum temperatures and diurnal temperature range on a dusty day and the nearest non-dusty day during June at Jodhpur are presented for the period 1968 to 1980 (Table 1). The data clearly show that on a dust storm day the diurnal temperature range is curtailed as a result of lower maximum and higher minimum temperatures. The mean temperature differences between dusty and non-dusty days were observed to be 2.8° C and 3.2° C for May and June respectively.

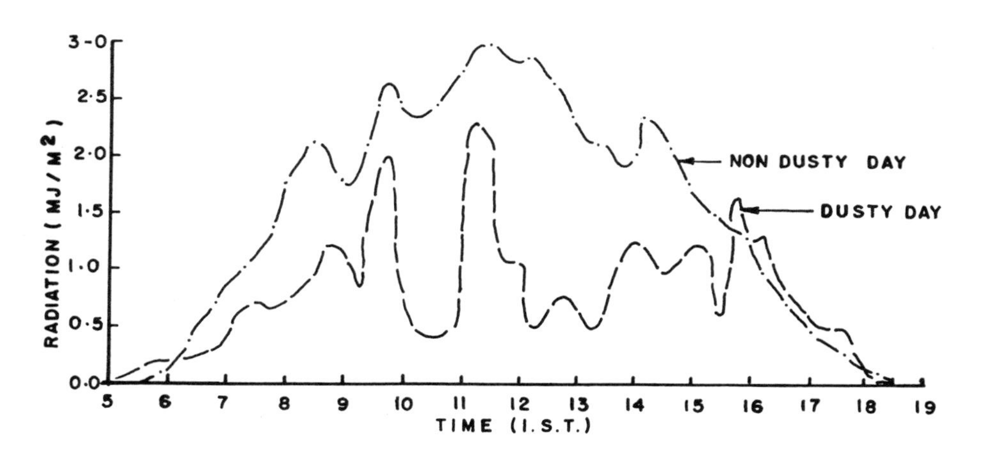

Figure 1: *Incoming solar radiation on dusty and non-dusty days.*

TABLE 1: *Influence of dust on diurnal temperature range at Jodhpur.*

Year	Dust storm day Maximum °C	Dust storm day Minimum °C	Dust storm day Range °C	Non dust storm day Maximum °C	Non dust storm day Minimum °C	Non dust storm day Range °C
1968	*40.6	26.5	14.1	42.7	26.7	16.0
1969	41.5	29.6	11.9	41.1	26.4	14.7
1970	39.5	27.4	12.1	40.3	26.8	13.5
1971	39.3	27.9	11.5	39.4	25.6	13.8
1972	39.7	28.2	11.5	39.9	20.3	19.6
1973	40.7	27.8	12.9	40.9	23.7	17.2
1974	36.4	26.7	9.7	36.5	22.6	13.9
1975	40.3	29.7	10.6	39.7	26.1	13.6
1976	41.5	28.9	12.6	42.0	25.3	16.7
1977	40.1	29.6	10.5	42.0	22.5	19.5
1978	43.3	28.4	14.9	43.3	24.6	18.7
1979	39.7	28.0	11.7	42.7	26.6	16.1
1980	41.9	30.8	11.1	43.4	26.3	17.1
Mean			13.0			16.2

* Day of dust raising winds.

INFLUENCE ON EVAPORATIVE DEMAND OF THE ATMOSPHERE

Dust storm activity, which is always associated with a strong wind regime, increases the evaporative demand of the atmosphere in the arid regions through advection and can result in increased water loss from water bodies and agricultural crops. Analysis of open pan evaporation data during dust storm days and non-dust storm days in June from 1968 to 1980 revealed that the mean open pan evaporation during a dusty day (17.4 mm/day) was 18% more than that during a subsequent non-dusty day.

This can be attributed to the fact that even though the dust activity results in decreased energy input, net radiation tending to decrease the evaporative losses, the influence of the aerodynamic factors associated with the strong wind regime dominates, increasing the evaporative losses.

CONTROL OF DUST MOVEMENT AND WIND EROSION

Dust movement and wind erosion are closely associated with the wind velocity and the erodibility of the soil fractions besides soil moisture status and vegetation cover. Mann (1980) classified the methods of control of dust movement and wind erosion into three main groups.

(a) Those which remove the abrasive material from the wind,

(b) Those which reduce the velocity of the wind near the soil surface, and

(c) Those which reduce the erodibility of the soil.

Of the various methods that fall under these three categories, shelter belt plantation, strip cropping, stubble mulch farming, proper tillage practices, sand dune stabilization with grasses and trees etc. hold promise for reducing the dust content and wind erosion in the arid regions of N.W.India.

ACKNOWLEDGEMENTS

The authors are grateful to Dr. K.A. Shankarnarayan, Director, CAZRI, for providing necessary facilities for carrying out the above study.

REFERENCES

BRYSON, R.A. (1971): Climatic modification by air pollution. Paper presented at Int. Conf. on Environ. Future, Helsinki, Finland 27 June-3 July, 1971, pp. 16.

GOUDIE, A.S. (1978): Dust storms and their Geomorphological implications. J. Arid Environ. 1: 291-310.

JUNGE, C. (1979): The importance of mineral dust as an atmospheric constituent. In: Sharan dust. C. Morales (Ed.), Scope, 14, John Wiley and Sons, New York, pp. 43-60.

MANN, H.S. (1979): Dust storms. Sci.Rep. 44-48.

MANN, H.S. (1980): Soil erosion and sand movement techniques for control. FAO/DANIDA Training course on sand dune stabilization, shelter belts and afforestation in dry zones, March 3-30, 1980 lecture notes. pp. 27.

MANN, H.S. and RAMAKRISHNA, Y.S. (1980): Dust and dust storms in the Indian desert. Int. workshop on physics of desertification, Trieste, Italy, Nov. 10-18, 1980, pp. 29.

RAMAKRISHNA, Y.S., SASTRI, A.S.R.A.S. and RAMANA RAO, B.V. (1983): A note on the prediction of annual dust storm activity over W. Rajasthan (communicated to J. Soil and Water Conservation).

VISHNU MITTRE (1977): Origin and history of the Rajasthan desert-Palaeobotanical evidence. In: Desertification and its control. P.L. Jaiswal (Ed.), I.C.A.R., New Delhi, pp. 6-9.

URBAN CLIMATIC CHANGES AND THEIR IMPACT ON HUMAN COMFORT AT DELHI

B. Padmanabhamurty
(Meteorological Office, New Delhi-3)

Abstract: - The consequences of urbanization on the climatic parameters, viz. temperature, humidity, wind speed, radiation, evaporation, precipitation and water balance have been examined and their impact on human comfort at Delhi are discussed.

INTRODUCTION

Urban climatic changes are caused by the increased surface roughness, the changed albedo and heat storage capacities resulting from the replacement of greenery by concrete buildings. The relative warmth of large cities, known as the 'urban heat island effect', is a well documented example of such local effects. It has been established beyond reasonable doubt that urban agglomerations cause measurable changes in the atmosphere immediately adjacent to them. Temperatures are increased, horizontal winds are slowed and updrafts induced, turbulence and cloud formation are increased and surface humidities are reduced. Most apparent is the increase in pollutants. They reduce solar radiation intensity and shorten sunshine duration. Their effect on cloud formation and rainfall over and in the vicinity of the cities is still somewhat uncertain but evidence points to occasional cases of stimulation of precipitation and perhaps some rare cases of inhibition.

Although city pollutant plumes have occasionally been followed for distances of several hundred kilometers, there is presently no sign of other notable effects of city influence on meteorological variables beyond a few or utmost several tens of kilometers. But as cities grow into large conturbations one can foresee that they will have notable regional weather effects.

HEAT ISLANDS

To study the effect of urbanization on meteorological parameters, a network of observatories needs to be established. In the absence of a close network of urban climatological observatories, mobile temperature surveys can be conducted. An isothermal analysis of observations from such a network and mobile survey at Delhi points out the existence of warm pockets and cold pools (Bahl and Padmanabhamurty, 1979). The intensity, size, shape and position of warm pockets have been studied with reference to the wind speed and direction.

The temperature excess of the urban areas over rural surroundings in the case of Delhi in the early morning on individual nights during cold weather period is of the order of 5 - 7° C.

From the network of urban climatological observatories in Delhi the mean daily heat island intensities at maximum and minimum epochs in different months are given in Table 1 (Padmanabhamurty and Bahl, in press).

WIND FIELD

Urban climate is influenced considerably by wind. When winds are light, near surface speeds are greater in the built-up area than outside, whereas the reverse relationship exists when the winds are strong. In winds of less than 4m/s, there was a 20% increase in speed over the city, the greatest increase being in winds of less than 1.3m/s. Evidence exists of cyclonic curvature of wind over urban areas in strong winds and anticyclonic curvature in light wind (Padmanabhamurty and Bahl, 1981).

HUMIDITY FIELD

Relative humidities, being a function of prevailing temperature, are normally found to be inversely related, in towns, to the local intensity of the urban heat island. On an average urban-rural differences are reported to be 5% but on individual nights the difference may approach 20%-30% (Padmanabhamurty and Bahl, 1982).

TABLE 1: *Mean heat island intensities (°C) at (a) maximum and (b) minimum temperatures epoch at Delhi.*

Month	Heat island intensity (°C)		Month	Heat island intensity (°C)	
	Max.	Min.		Max.	Min.
Jan	4	6	Jul	2	5
Feb	3	4	Aug	2	4
Mar	4	6	Sept	2	3
Apr	3	6	Oct	4	6
May	4	5	Nov.	2	6
Jun	2	3	Dec	3	6

TOTAL RADIATION

Urban-rural radiation differences in winter at Delhi point out that urban radiation is less than rural radiation which could be attributed to comparatively higher pollution levels in urban Delhi over rural Delhi. Occasions were there when rural radiation was less than urban radiation which may be due to change of wind direction resulting in transport of pollutant from urban complexes to rural locations (Padmanabhamurty and Mandal, 1982).

PRECIPITATION

Reports are divergent in literature on the effects of urbanization and industrialization on precipitation. Arguments advanced are that large cities with super-abundance of condensation nuclei, influence precipitation processes towards a reduction in precipitation with increasing small droplets in the form of clouds and smog, while the contribution of smaller urban complexes may supply the right amount of additional condensation nuclei to increase precipitation amounts.

Isohyetal distribution at Delhi from June to September showed pockets of higher rainfall in all months but being intense in July/August. These pockets of higher rainfall correspond to the congested urban agglomeration supporting the hypothesis that urbanization leads to increased buoyancy and convection resulting in increased precipitation (Padmanabhamurty and Bahl, in press).

EVAPORATION

Urban-rural evaporation differences during two years point out that rural evaporation is more than urban which is in tune with higher radiation at the former than at the latter place.

WATER BALANCES

In view of the inadvertent modification brought about by urbanization on the meteorological parameters, climatic water balances were computed for urban and rural locations of Delhi. In these computations the heat island effect both during day and night and the variation in precipitation at the region of heat island and rural areas were given due consideration. The resultant water balances are shown in Fig. 1. Urban Delhi has more precipitation and water surplus compared to rural areas. This has resulted in a moister climate at urban Delhi as compared to rural Delhi. According to Thornthwaite's (1955) climatic

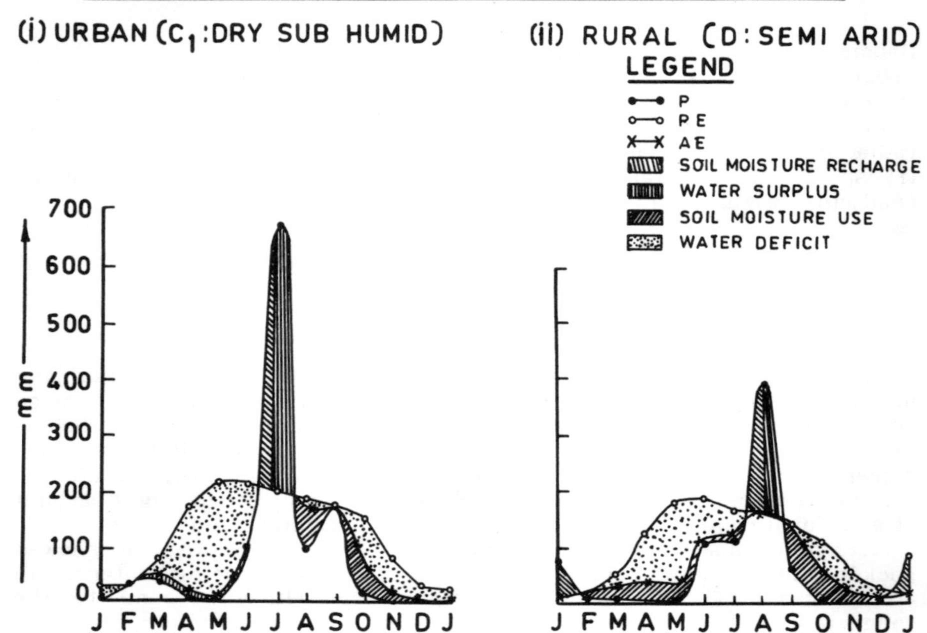

Figure 1: *Climatic water balance diagrams of urban and rural New Delhi. Note that the urban area appears slightly more humid than the rural area.*

classification by moisture regime, urban Delhi falls under Dry sub-humid type while rural Delhi remains under semi-arid category. These differences in climatic type could contribute to different conditions of comfort at these two locations.

HUMAN COMFORT AT DELHI

The effect of heat island on the comfort or discomfort in Delhi and its surrounding areas have been examined. Day-time conditions show that February, March, November and December are comfortable at rural Delhi, but in urban Delhi January, February and December are comfortable. The heat island in January enabled urban Delhi to be comfortable but rural Delhi remains uncomfortable on the colder side. On the warmer side from April to October rural Delhi is uncomfortable but urban Delhi is uncomfortable from March to November.

Considering night-time conditions, rural Delhi is comfortable in May and June only. July, August, September are uncomfortable on warmer (humid) side and January to April and October to December are uncomfortable on the colder side. At urban Delhi, because of the heat island, April, May and October are comfortable. June to September and October to December, January to March are uncomfortable on the warmer (humid) and colder sides respectively.

CONCLUSIONS

Inadvertent modification of local climate of urban complexes are likely to result in significant changes in meteorological parameters. Some of the changes, particularly the heat island effect, could be taken advantage of in air conditioning and refrigeration. However, the adverse effects, for example, transport of pollutants from the periphery of urban areas downtown, higher precipitation, quick runoff etc., outweigh the advantages. It is, therefore, desirable that in planning urban complexes, built up areas and concrete surfaces are to be interspersed with wide lawns and parks to mitigate the formation of intense heat islands.

REFERENCES

BAHL, H.D. and PADMANABHAMURTY, B. (1979): 'Heat island studies at Delhi', Mausam. 30: 119-122.

PADMANABHAMURTY, B. and BAHL, H.D. (1981): 'Ecoclimatic modification of Delhi due to urbanization'. Mausam, 32: 295-300.

PADMANABHAMURTY, B. and BAHL, H.D. (1982): 'Some physical features of heat and humidity islands at Delhi'. Mausam, 33: 2, 211-216.

PADMANABHAMURTY, B. and MANDAL, B.B. (1982): 'Urban-rural radiation differences'. Mausam, 33, 4, 509.

PADMANABHAMURTY, B. and BAHL, H.D. (In press): 'Isothermal and Isohyetal patterns at Delhi as a consequence of urbanization'. Mausam.

THORNTHWAITE, C.W. and MATHER, J.R. (1955): 'Water Balance of the Easter'. Publications in Climatology, Drexel Inst. of Tech. Vol. 8, No. 1.

ROLE OF SHELTER BELTS IN ARID SITUATIONS

G.G.S.N. Rao, B.V. Ramana Rao and Y.S. Rama Krishna
(Central Arid Zone Research Institute, Jodhpur)

Abstract: - A comparative study of three row tree shelter belts of *Prosopis juliflora*, *Cassia siamea* and *Acacia tortilis*, conducted during 1977 to 1982 at the Central Arid Zone Research Institute, Jodhpur, revealed that at 2H distance, *Cassia siamea* was efficient in wind speed reduction. However, at larger distances wind speeds were effectively reduced by *Acacia tortilis*. Tree shelter belts resulted in a decrease of 8 to 12% in evaporation rates. Micro crop shelter belts made up of tall growing pearl millet crop resulted in improving microclimate and in increasing vegetable yields of cowpea and lady's finger crops by 21 and 41% respectively.

INTRODUCTION

Apart from scarcity and high variability of rainfall and low fertility of soils, agricultural production in the arid zone of India is also considerably influenced by wind erosion due to prevalence of strong wind regime during summer season. As this region will be practically barren during summer, the damage caused by these strong winds is manyfold. Such climatic conditions in the Indian arid zone set a limit to the production which can hardly support both human and animal population, thus leaving a wide gap between production and demand.

A solution that has found worldwide acceptance to the problem of strong wind regime is the establishment of shelter belts. The primary effect of natural or artificial shelter belts is the reduction of wind speed. However, this objective often takes a second place because of the interest in the secondary effects, for example, changes in microclimate influencing the growth and productivity of the sheltered crop.

IMPORTANCE OF SHELTER BELTS IN ARID REGIONS

Shelter belts are erected in many parts of the world mainly to improve the living conditions for human beings and livestock and to provide a favourable micro-climate for the growing of plants and to minimise the problems of soil erosion. The major role of shelter belts in arid regions is to provide protection to crops from hot winds, to prevent erosion of fertile soils and to arrest the sand drift which often results in chocking of canals, blocking of rail tracks and roads. In arid areas of India farmers are aware of the usefulness of shelter belts, and fields protected by a primitive type of wind breaks in arid regions of Rajasthan, locally known as "Matts", were found to increase yields considerably over crops grown on unprotected fields (Jodha, 1967). With energy crises, the importance of plantation of the shelter belts in arid regions of India is increasingly realised because of their utility as additional sources of fodder and fuel.

STUDIES ON SHELTER BELTS IN INDIAN ARID REGION

Studies on shelter belt plantations were initiated at Central Arid Zone Research Institute (CAZRI), Jodhpur, way back in 1960. Successful plantations of wind breaks were carried out at Central Mechanised Farm in Suratgarh and on the sides of railway tracks with plant species like *Acacia nilotica,* sp. *Indica* and *Dalbergia sisso* (Bhimaya and Chowdhary, 1961) by the CAZRI Scientists. Three row tree shelter belts with plant species *Acacia tortilis, Cassia siamea* and *Prosopis juliflora* as side rows and *Albezzia lebbek, Azadirachta indica* and *Tamarix articulata* as Central rows were planted in 1973 at Central Research Farm of CAZRI, Jodhpur. Observations on wind speed reduction at different distances in the leeward side, temperature profiles and measurements on evaporation using open pan evaporimeters were recorded during summer and monsoon seasons for the years 1977 to 1979. Vertical profiles of wind speeds were also recorded during 1980 to evaluate these three prominent plant species of the desertic regions.

RESULTS AND DISCUSSION

WIND REGIME AND EVAPORATIVE DEMAND AS INFLUENCED BY SHELTER BELTS

From the wind speed observations, the percentage of reduction in wind speed measured daily at 1m above ground level in the leeward side at 2H*, 5H and 10H and in the windward side by different shelter belts were worked out for each year. Data collected during 1977-1979 was averaged and presented in Table 1.

It is seen that wind speed reduction was higher during monsoon season than during summer season due to better canopy growth.

* H is the average height of shelter belt.

TABLE 1: *Percentage reduction of mean wind speed at different distances in the leeward side (1977-1979).*

Shelter belt	Summer season Distance from the shelter belt			Monsoon season Distance from the shelter belt		
	2H	5H	10H	2H	5H	10H
Prosopis juliflora	33	17	12	38	26	21
Cassia siamea	36	17	13	46	36	24
Acacia tortilis	36	25	13	46	36	20

Performance of *Cassia siamea* and *Acacia tortilis* in reducing wind speeds was better than *Prosopis juliflora* at distance 2H and 5H, while at the larger distance of 10H, the difference between species has not varied much.

The predominant wind directions are from South-Southwest (SSW), Southwest (SW) and West Southwest (WSW) in these regions. Wind speed reduction by different shelter belts in the above three directions were calculated. It was observed that the highest reduction in wind speed was observed by *Cassia siamea* shelter belt at a distance 2H, when the wind direction was between SSW and WSW. At greater distances of 5H and 10H and for wind speeds greater than 10 km/h, wind speed reduction by *Acacia tortilis* was highest followed by *Cassia siamea* for all the above mentioned wind directions. At lower wind speeds, less than 10 km/h, for the above mentioned distances on the leeward side, the reduction in wind speed by *Cassia siamea* was highest compared with the other two shelter belts. It was found that at a distance 2H for all wind speeds, *Cassia siamea* is more effective in reducing wind speeds and at larger distance and for higher wind speeds, *Acacia tortilis* is found to be better. This may be due to the difference in shape and structure of the plant canopies.

Evapotranspiration (PE) in arid regions is greatly controlled by wind speeds. Krishnan and Kushwaha (1971) showed that the correlations between the evaporation and aerodynamic and energy balance term of Penman's equation are 0.86 and 0.36 respectively. This clearly shows the influence of aerodynamic term on the process of evapotranspiration compared to energy balance term. Hence evapotranspiration rates by using Penman's (1948) equation at different distances in the leeward side was calculated by substituting the wind speed values recorded at the respective distances. As there was not much difference in the temperature between windward and leeward side distances, it is assumed to be same at all the sites of measurement and accordingly the PE values have been estimated at different distances in the leeward side. As pan evaporation data are available for a single year (1978), the values of pan evaporation and the PE values obtained at windward and at different distances in the leeward side are presented in Table 2.

TABLE 2: *Influence on evaporation by shelter belts.*

Month	Pan evaporation mm/day Wind-ward	Pan evaporation mm/day Lee-ward	Estimated values of PE (mm/day) in the leeward side									
			Wind ward	*Cassia Siamea*			*Prosopis juliflora*			*Acacia tortilis*		
				2H	5H	10H	2H	5H	10H	2H	5H	10H
April	11.0	9.5 (14)	6.6	6.0 (10)	6.1 (8)	6.2 (6)	5.9 (11)	6.1 (8)	6.4 (5)	5.8 (12)	6.2 (6)	6.5 (2)
May	14.5	13.4 (8)	8.7	7.9 (10)	8.1 (7)	8.3 (5)	7.9 (10)	8.0 (8)	8.3 (5)	7.7 (11)	7.9 (10)	8.3 (5)
June	13.4	12.4 (8)	8.2	7.5 (9)	7.7 (6)	7.9 (4)	7.5 (9)	7.6 (7)	7.9 (4)	7.3 (11)	7.5 (9)	7.7 (6)
July	6.0	5.7 (5)	5.0	4.6 (8)	4.7 (6)	4.8 (4)	-	-	-	4.6 (8)	4.7 (6)	4.8 (4)

* Figures in brackets indicate the percentage of reduction.

The data in Table 2 show that a decrease of 8 to 12% in evapotranspiration rate was observed on the leeward side at 2H distance by all shelter belts. At distances 5H and 10H the decrease was of the order of 2 to 6%. The reduction in evaporation demand as influenced by *Acacia tortilis* was slightly higher at all distances as compared to the other two shelter belts. The decrease of 8 to 14% in pan evaporation at a distance 2H on the leeward side of *Cassia siamea* is quite comparable in magnitude to PE estimated for the same distance.

MICRO-CROP SHELTER BELTS

The micro-crop shelter belts are essentially shelter belts made up of tall growing plant species, providing shelter to low growing plants like vegetable crops or any cash crops. Often such micro-crop shelter belts indicate added advantages with respect to improving the micro-climatic conditions in the sheltered regions, thereby improving the productivity and moisture use efficiency of the sheltered crop.

Much work has been done at CAZRI on influence of micro-crop shelter belts on vegetable production viz. cowpea and lady's finger during summer for the years (1976-1981) and (1976-1978) respectively. The wind barrier used in these studies was made up of three rows of pearlmillet sown perpendicular to the normal wind regime. The micro-crop shelter belt maintained lower air and soil temperature, higher humidities and lower vapour pressure deficit and lower wind regime throughout the active growth phase in the sheltered area. It was found instrumental to increase vegetable yields to the order of 21 and 44% in case of cowpea and lady's finger respectively (Table 3).

TABLE 3: *Vegetable yield (q/ ha) of lady's finger and cowpea as influenced by micro-crop shelter belt.*

Treatment	1976	1977	1978	1979	1980	1981	Mean yield	Percentage increase in yield due to shelter
Lady's finger (76-78)								
Unshelter	24.3	17.6	28.7	-	-	-	23.5	
Sheltered	48.2	22.2	29.3	-	-	-	33.2	41
Cowpea (76-81)								
Unshelter	30.2	14.8	52.0	15.5	23.4	35.8	28.6	
Sheltered	41.1	24.7	56.4	17.9	26.9	40.8	34.6	21

In addition to increase in vegetable yields, the micro-crop shelter belt also provided additional remuneration through the yield of pearlmillet fodder, the total additional remuneration from the system working out to be Rs.1600/ha. It had also reduced the water use by the crops and increased the vegetable yields per unit of water, making thereby the best use of limited water resources available in the arid regions of India.

ACKNOWLEDGEMENTS

The authors are grateful to Dr. K.A. Shankaranarayan, Director, Central Arid Zone Research Institute, Jodhpur for his encouragement and guidance in carrying out the above study.

REFERENCES

BHIMAYA, C.P. and CHOWDHARY, M.D. (1961): Plantation of wind break in the Central Mechanised Farm, Suratgarh - an appraisal of technique and results. Indian Farming 87: 354-367.

JODHA, N.S. (1967): Capital formation in arid agriculture. Ph.D. thesis, University of Jodhpur, Jodhpur.

KRISHNAN, A. and KUSHWAHA, R.S. (1971): A critical study of evaporation by Penman's method during the growing season of vegetation in the arid zone of India. Arch. Met. Geoph. Biokl. Ser. B. 19: 267-276.

PENMAN, H.L. (1948): Natural evaporation from open water, bare soil and grasses. Proc.Roy.Soc. (A) 193: 120-145.

ACIDITY AND TRACE METAL CONCENTRATIONS IN RAIN WATER OVER SOME PARTS OF INDIA AND THEIR SIGNIFICANCE TO TERRESTRIAL AND AQUATIC LIFE

B.K. Handa
(Central Chemical Laboratory, 4, Sapru Marg, Lucknow-226001)

Abstract: - The studies carried out during the last 16 years on the chemical composition of rain water in different parts of India have revealed that the concentration of various constituents shows both temporal and spatial variations. The spectre of acid rain is likely to be confined to big industrial towns like Calcutta, Bombay etc. In Calcutta, for example, rain water having pH 4.40 was found to occur. The concentration of heavy metals like Zn, Pb, Mn, Mi, Fe, Cd and Cu were generally much below those reported for rain water in the more industrially developed countries of the world.

INTRODUCTION

The pH of rain water is affected both by natural as well as anthropogenic sources. The pH of rain water in equilibrium with the partial pressure of CO_2 in the atmosphere should be around 5.7. However, due to emission of sulphur gases and oxides of nitrogen, which subsequently get oxidised to sulphuric and nitric acids respectively, the pH of rain water in certain industrial areas may fall below this value, giving rise to 'acid' rain. In fact, while weakly acidic rain may be beneficial to plant life, the relatively strong acid rain (pH below 5.0) may affect the foliage of trees and plants and may even adversely affect their growth activities by interacting with soil minerals in the root zone by leaching away the nutrient elements, or fix them in non-available form or release some constituents from soil minerals which may become toxic to plants. Even aquatic biota may be affected.

The concentrations of trace metals in the atmosphere have become a matter of considerable concern for many reasons, the most important being:

a. metals in air may be health hazard to human beings and animals;
b. these metal ions may act as catalysts in atmospheric reactions, some of which cause production of acids viz. oxidation of sulphur and nitrogen compounds;
c. these metal ions may promote corrosion of materials;
d. may cause far reaching changes in the receiving eco-system and
e. in association with suspended particulate matter may induce changes in climate at micro or even at macro levels.

For the last several years studies on the chemical composition of rain water in different parts of India have been initiated by the author and his associates (Banerjee et al., 1967; Bhatia et al. 1976; Handa 1968, 1969 a,b,c,d, 1971, 1973 a,b, 1978, 1979, 1982; Handa et al. 1983; Kumar et al. 1979 etc). The present study is an extension of this work, to compare the chemical composition of rain water in Lucknow in 1982, with the data obtained earlier.

MATERIALS AND METHODS

Rain water samples were collected in polyethylene containers fitted with large diameter polyethylene bottles. These containers were placed on a raised platform to prevent contamination of the collected sample from splashings from the ground. Immediately after the rain had ceased, the sample was analysed for pH and other constituents like, EC, HCO_3, Cl, NO_3, Ca, Mg, Na, K, SiO_2 by standard methods described in the literature. An aliquot of the rain water sample was also acidified with reagent-grade HNO_3 and concentrated 50 times at low heat. This concentrated sample was analysed for Li, Ag, Cu, Mn, Fe, Pb, Cr, Ni, Co, Cd, Me, Sr, Rb, Cs and Zn using a Perkin Elmer Model 306 atomic absorption spectrophotometer (Handa, 1978).

RESULTS AND DISCUSSION

The chemical composition of rain water is affected by several factors. In addition to natural sources, a great variety of technical and industrial processes such as metal smelters, blast furnaces, thermal power plants, petrochemical plants, chemical plants etc. contribute a wide spectrum of metals to the atmospheric aerosols. The combustion of fossil fuels, burning of garbage, motor traffic exhaust, spraying (aerial) of insecticides, forest fires, volcanic activities, and bacterial reduction in stagnant water bodies are other important sources of gaseous and metallic constituents of the atmosphere. The output from these sources is varied and more often than not, discontinuous. Thus concentrations of these constituents, both in gaseous and particulate form, vary considerably in the atmosphere, and thus the concentration of these constituents in rain water also varies. For example, a considerable portion of iron, copper, lead and zinc emissions come from point sources like smelters, incinerators, blast furnaces etc. The contribution of each source to the atmospheric burden of these elements depends not only on the quantity and physical characteristics of the aerosols containing these metallic constituents but also on the precise point of aerosol release. The location and height of the emitter, topography of the adjacent area - whether flat or hilly, - meteorological conditions (dry, windy or rainy etc.) and aerodynamic sizes of the aerosols determine whether these metals will be dispersed widely or concentrated in a particular area. In addition to the influence of these sources, the concentration of these metallic constituents in the atmosphere will be affected by entrainment of local soils, exudates from plants, the import of these metals from distal sources and the ageing history of the aerosol particles. Meteorological factors such as wind direction, trajectory of the precipitating mass, intensity, frequency and total quantity of rainfall, the height of the cloud base etc. will affect the 'rain-out' or 'wash-out' efficiency and thus the concentrations of these constituents in rain water. Thus the chemical composition of rain water may show considerable temporal and spatial variations. For example, areas located near the emitter sources, viz. near foundaries, smelters etc., may have a relatively high concentration of these constituents in rain water (and may even show diurnal, daily, weekly or monthly variations) as compared to rain water sampling sites which are located far away from these surfaces.

Similarly the release of sulphur gases by industries, burning of fossil fuels, through bacterial reduction of sulphate in stagnant water bodies or release of oxides of nitrogen to the atmosphere by various sources can affect the concentrations of sand N species in rain water as well as the pH of rain water. In Lucknow the pH of rain water was found to be 7.31. In the Doon Valley of northern U.P. lying near the foot of the Himalayas, the pH of rain water samples averaged 5.97. It is interesting to record that the average sulphate content of rain water samples in Calcutta was found to be the highest, followed by Dehradun rain water. Apparently in these places sulphur gases either from industrial waste emissions or from stagnant water bodies or from natural sources which are being introduced into the atmosphere, get oxidized to sulphate ions simultaneously.

PHENOMENON OF ACID RAIN

The examination of the data given in Tables 1 and 2 is in accordance with the hypothesis outlined above. In Calcutta, which is an industrial city, the average pH value for rain water was found to be only 5.21, while in Bankipur village, which is far (about 300 km north of Calcutta) from any industrial activity, the average pH of rain water samples was 6.10. It may be recorded that the pH of pure rain water in equilibrium with partial pressure of CO_2 in the atmosphere should lie around 5.7. The average pH of rain water in Chandigarh (NW India) was found to be 6.02, while in Lucknow, which lies in the heart of the Indo-Gangetic plain, the average pH of rain water was found to be 7.31. In the Doon Valley of northern U.P. lying near the foot of the Himalayas, the pH of rain water samples averaged 5.97. It is interesting to record that the average sulphate content of rain water samples in Calcutta was found to be the highest, followed by Dehradun rain water. Apparently in these places sulphur gases either from industrial waste emissions or from stagnant water bodies or from natural sources which are being introduced into the atmosphere, get oxidized to sulphate ions lowering the pH of the rain water.

The data in Table 3 give the chemical composition of some acid rain

TABLE 1: *Chemical compisition of rain water in some parts of India.*

Parameter	Sampling site									
	Calcutta		Chandigarh		Bankipur		Dehradun		Lucknow	
	Av	SD	Av	SD	Av	SD	Av	SD	Av	SD
pH	5.21	0.74	6.02	0.47	6.10	0.42	5.97	0.23	7.31	0.45
EC	-	-	-	-	-	-	-	-	13.8	10.8
HCO_3	1.97	2.55	2.96	2.95	2.21	1.64	13.1	11.9	6.6	5.9
Cl	2.11	0.71	0.34	0.43	0.61	0.32	0.43	0.33	1.8	1.7
SO_4	3.33	1.81	0.23	0.15	0.44	0.16	1.87	0.71	0.5	1.2
NO_3	0.42	0.68	0.19	0.11	0.38	0.15	-	-	0.5	0.87
F	0.08	0.04	-	-	-	-	-	-	0.05	0.02
Ca	1.81	2.93	0.69	0.77	0.94	0.64	3.39	3.49	1.47	2.39
Mg	0.06	0.08	0.12	0.22	0.15	0.15	0.64	0.71	0.61	0.39
Na	1.16	0.42	0.07	0.09	0.31	0.15	0.74	0.47	0.87	0.50
K	0.32	0.68	0.09	0.09	0.11	0.06	0.21	0.11	0.57	0.80
SiO_2	-	-	0.10	0.05	0.44	0.32	-	-	1.09	0.30

Note: Concentrations are in mg/l; EC values in microsiemens/cm at 25° C Av = Average SD = standard deviation

Period of collection: Calcutta Sept. 1967 to March 1968
Chandigarh Aug. Sept. 1971
Bankipur June to Oct. 1968
Dehradun March to July 1967
Lucknow January to Dec. 1978

TABLE 2: *Monthly variations in chemical composition of rain water over Calcutta.*

Parameter	1967					1968					
	June	July	Aug.	Sept.	Nov.	Jan.	Feb.	March	Apr.	May	June
pH	6.20	6.05	6.13	5.11	6.10	6.10	6.05	7.15	6.35	6.68	6.89
HCO_3	14.40	7.16	5.65	1.19	8.00	4.00	7.00	6.00	12.70	16.80	4.80
Cl	4.42	3.13	2.24	1.04	3.00	2.50	2.00	2.00	6.10	9.80	2.59
SO_4	5.14	5.54	3.16	2.59	-	9.66	4.50	4.90	5.20	5.69	0.98
NO_3	0.10	0.22	0.28	0.32	1.60	-	0.06	0.11	1.57	0.67	0.51
Ca	5.94	3.76	3.44	0.81	11.60	8.00	3.60	3.20	8.88	7.29	2.29
Mg	1.08	0.43	0.57	0.09	tr	0.12	tr	tr	1.02	1.61	0.69
Na	1.32	1.66	0.87	1.13	1.30	0.30	1.32	1.70	3.54	7.09	1.17
K	0.38	0.49	0.43	0.14	3.10	0.10	0.90	0.50	1.38	1.87	0.61
F	-	-	-	0.07	-	-	-	-	-	0.12	0.02
SiO_2	-	-	2.27	1.37					2.07	0.85	1.20

water samples from Calcutta analysed by the author in 1967. It would be seen that the lowest pH value recorded was 4.40. The occurrence of relatively high concentrations of sulphate ions indicates that the main cause of lowering of pH of rain water is to be attributed to the

TABLE 3: *Chemical composition of 'acid' rain over Calcutta (Handa, 1969).*

Date of collection	pH	Na	K	Ca	Mg	HCO_3	Cl	SO_4	NO_3	F	D	
9th Sept.'67	4.80	1.31	0.14	3.00	0.18	4.15	1.50	3.59	1.90	0.10	0.002	a
"	5.20	0.47	0.03	1.60	0.24	3.00	1.00	3.60	-	-	0.007	b
"	5.10	0.92	tr	2.40	0.24	4.00	1.00	4.20	-	-	-	c
10th Sept.'67	5.00	1.68	0.13	0.60	0.02	1.00	0.75	3.00	0.16	-	0.012	a
"	4.95	1.15	0.02	0.40	tr	1.00	0.50	2.11	-	-	0.036	b
"	4.80	-	-	0.30	0.01	1.00	1.00	2.80	0.04	0.12	0.014	c
11th Sept.'67	4.85	1.00	0.16	0.40	0.06	0.50	0.75	2.40	0.02	0.08	0.012	a
"	4.85	1.90	0.05	0.40	tr	1.00	1.00	2.60	0.12	-	0.010	b
"	4.85	1.16	0.20	0.40	0.12	0.75	0.50	2.80	-	-	0.014	c
"	5.35	1.69	0.10	0.40	0.06	tr	1.50	3.02	-	-	0.002	d
"	4.45	0.95	0.13	0.32	0.01	tr	0.50	1.50	0.05	0.10	tr	e
"	5.05	1.24	0.10	0.24	tr	tr	0.50	2.41	0.04	-	0.002	f
"	4.60	1.39	0.20	0.20	0.07	tr	0.75	2.77	-	0.04	-	g
"	4.40	1.31	0.16	0.20	0.08	tr	0.50	2.48	-	0.01	0.002	h
"	4.80	0.96	0.20	0.20	tr	tr	0.63	-	-	-	0.002	i
"	4.50	0.63	0.16	0.28	0.02	tr	0.75	-	-	-	0.002	j

Note: Concentrations are in mg/l
a, b, c, d .....j represent successive showers on the same day.

oxidation of sulphur gases (produced by industrial waste emissions, fossil fuel burning, bacterial reduction of sulphate in stagnant water bodies etc.) to sulphuric acid. It must, however, be recorded that not in all the cases the pH of rain water over Calcutta was found to be below 5.5, but in some cases pH values above 6.0 were also recorded (Handa, 1969 a,b,c,d).

SPATIAL VARIATIONS IN CHEMICAL COMPOSITION OF RAIN WATERS

The data given in Table 1 show clearly that the chemical composition of rain water varies from place to place, depending on various factors like nearness to sea, topography of the area and contributions from anthropogenic sources etc. It is interesting to record that the chloride content of rain water samples from Lucknow was found to be higher than that for Calcutta, which indicates considerable contribution from localized sources. However, the sulphate content in rain water samples over Calcutta was found to be the highest for reasons which have already been discussed in the preceding paragraphs.

TEMPORAL VARIATIONS IN CHEMICAL COMPOSITON OF RAIN WATER

In Table 2, monthly variations in chemical composition of rain water have been given. The data are in conformity with the hypothesis outlined earlier and show that the chemical composition of rain water at any one place may vary considerably from time to time.

Table 4 shows average annual concentrations/values of various parameters for rain water samples over Lucknow. The examination of the data shows the following trends:

a. the EC of the rain water samples seems to be increasing as is shown by the increase in the average EC values of rain water samples;
b. the bicarbonate content, with some variations, also seems to be increasing;
c. the calcium ion concentration also seems to be increasing but the concentration of other constituents like Cl, Na, K etc. is fluctuating a lot; with the result no clear trend was observed.

The data show the importance of local factors in affecting the chemical composition of rain water.

OCCURRENCE OF TRACE METALS IN RAIN WATER

The trace metals present in the atmosphere play an important role in causing certain diseases in man and his resource organisms. These effects may be direct or indirect. Amongst the direct effects one may mention the inhalation of Cr, Pb, Cd, Mn etc. bearing aerosol particles

TABLE 4: *Average values and standard deviation of some constituents present in rain water over Lucknow from 1978-1982.*

Parameter	1978		1979		1980		1981		1982	
	Av.	SD	Av.	SD	Av.	SD	Av.	SD	Av.	SD
pH	7.31	0.45	7.56	0.45	7.21	0.36	7.48	0.45	7.31	0.56
EC	13.8	10.8	23.72	8.05	25.1	22.2	30.5	27.5	38.1	42.2
HCO_3	6.6	5.91	9.22	1.77	8.83	11.5	11.3	11.3	15.1	18.0
Cl	1.78	1.66	1.33	0.56	1.18	1.01	0.93	0.77	2.14	2.37
SO_4	0.53	1.18	0.65	0.28	-	-	-	-	-	-
NO_3	0.52	0.87	0.33	0.18	1.96	2.31	1.26	1.45	1.28	1.34
Ca	1.47	2.39	1.35	1.07	1.70	3.52	3.34	4.05	3.74	5.70
Mg	0.61	0.39	0.86	0.53	0.43	1.27	0.14	0.42	0.42	1.17
Na	0.87	0.50	0.93	0.55	0.58	0.69	0.76	0.83	0.84	0.99
K	0.57	0.80	0.42	0.38	0.35	0.43	0.41	0.50	0.85	1.36
F	0.05	0.02	0.04	0.02	-	-	0.03	0.04	0.24	0.27
SiO_2	1.09	0.30	0.53	0.46	0.24	0.58	-	-	0.54	0.68

Concentrations in mg/l: EC in microsiemens/cm at 25° C.
SD = standard deviation Av. = average value.

which may cause serious damage to various human tissues via the 'lung burden'.

Amongst the indirect effects mention may be made of the adsorption of trace elements by the plants directly from the atmosphere and which are then carried in the food chain. These metallic constituents may also reach the land surface as 'dry fall-out' or with precipitation (as washout or rain-out process). These may then be retained by the clay particles or organic matter present in the soil, and then may be taken up by the plant roots. These plants may then be eaten by humans or animals, and if their concentration in high, may prove harmful to man and/or animal. While at present no exact relationship is known to relate trace metal concentration present in rain water to its concentration in the atmosphere aerosols, one can say only approximately that the higher the metal ion concentration in the aerosol, its concentration is likely to be higher in the rain water also, provided other conditions viz. intensity, trajectory of the precipitating mass, etc. remain constant. Nevertheless, the data given in Table 5 do show that varying amounts of various trace metals are present in rain water; and their concentrations show considerable variations with time. Amongst the trace constituents iron is most abundant followed by zinc, copper, strontium, nickel and manganese. Molybdenum ions could not be detected but lithium, cadmium, cobalt, silver, chromium, lead, rubidium, casium etc. are present in lesser amounts.

While it is not possible to compute the concentrations of these metallic constituents in the atmosphere and to determine their adverse effects on human beings, the concentrations of some of the

TABLE 5: *Trace element content in rain water over Lucknow (U.P.) during the period 1978-1982.*

Parameter (μg/l)	1978		1979		1980		1981		1982	
	Av.	SD	Av.	SD	Av.	SD	Av.	SD	Av.	SD
Ag	0.35	0.82	0.72	1.12	0.41	0.96	0.53	0.63	1.59	2.85
Cu	4.3	6.8	3.78	3.26	3.09	4.23			5.54	6.82
Zn	14.3	8.31	14.5	7.17	9.76	10.25	10.4	8.5	14.6	9.65
Co	1.66	0.76	0.27	0.50	0.60	0.64	0.23	0.17	0.20	0.47
Mo	0.00	-	0.0	-	1.16	2.69	0.0	-	0.0	-
Cd	0.36	0.22	0.07	0.14	0.19	0.55	0.0	-	0.03	0.11
Sr	1.88	2.83	4.13	4.50	3.68	3.45	4.83	4.74	3.43	4.01
Li	0.04	2.05	0.0	-	1.00	2.78	0.22	0.13	0.44	0.62
Rb	0.30	0.61	1.33	1.16	0.61	0.37	0.11	0.30	1.31	1.80
Cs	0.02	0.07	0.73	0.85	0.41	0.44	0.94	0.80	0.29	1.14
Fe	-	-	-	-	26.5	26.0	30.9	24.8	43.0	38.7
Mn	-	-	-	-	2.45	2.75	2.44	1.54	4.84	6.57
Cr	-	-	-	-	-	-	-	-	0.31	0.85
Pb	-	-	-	-	-	-	-	-	1.46	2.83
Ni	-	-	-	-	-	-	-	-	4.56	5.13

Concentrations are in μg/l : Av = average (mean) value
SD = Standard deviation.

constituents present in rain water are of direct influence on the aquatic biota. For example the limit for cadmium, copper, zinc and lead for sensitive species in fresh water has been laid down to be 0.2 μg Cd/l; 5 μg Cu/1; 30 μg Zn/l and 10-25 μg Pb/l.

The data in Table 5 show that the average Cd concentration in rain water in 1978 exceeded the limit of 0.2 μg/l. Similarly in 1981, the average copper concentration was nearly four times the recommended maximum limit of 5 μg/l. While it is true that some portion of these metallic constituents will be taken up by the suspended matter present in surface water bodies and some dilution may occur by mixing, the significance of regular monitoring for these trace elements can not be overemphasized.

COMPARISON WITH DATA FROM OTHER COUNTRIES

A comparison of the trace element concentrations in rain water over different parts of the world (Jefferies & Snyder, 1981 and Lantay & Mackenzie, 1979) is quite interesting. In the highly (developed) industrialized countries, the lead content is found to be much higher than that found in the rain water over Lucknow. In Hamilton, Canada, for example, the lead content averaged 180 μg/l, whereas in Lucknow the average value was only 1.46 μg/l. The copper content is also generally higher in industrialized countries, except that in 1981 when the copper content in rain water over Lucknow was found to be anamolously high and exceeded the average Cu concent in rain water over Sweden and Norway. The iron content in rain water over Lucknow is generally more than that in rain water over Sweden but is much lower than that in some other countries. In Germany, for example, the averages for rain at Heidelberg and Göttingen are 1750 and 1400 μg Fe/l, where as in Lucknow the average value is only 43 μg Fe/l. Similar comparisons may be made for other metallic constituents.

CONCLUSIONS

The discussions outlined above have shown that the occurrence of 'acid' rains is likely to be confined to big cities like Calcutta or near sulphur emission sources. The rain water composition shows both spatial and temporal variations. Cd and Cu were present in rain water over Lucknow, in amounts exceeding those recommended for sensitive aquatic species, thereby emphasising the importance of regular monitoring of rain water samples with respect to trace elements. Further, in most of the cases, the concentration of metallic constituents present in rain water over Lucknow was much lower than those present in rain water over the highly industrialized countries of Western Europe, Britain, Canada, Australia and U.S.A.

REFERENCES

BANERJEE, P., MUKHOPADHYAY, P.K., HANDA, B.K. and CHATTERJI, S.D. (1967): Record of an abnormal rainfall over Calcutta on 10th Oct. 1965. J. Met. & Geophys. 13: 1-2.

BHATIA, S.S., THAKUR, K.S., GAUMAT, M.M., KUMAR, A., SEHGAL, R.K. and HANDA, B.K. (1976): Chemical composition of rain water over Lucknow. Symp. Contribution of Earth Sciences towards Research & Developmental Activities in Northern Region, G.S.I. Lucknow, Nov. 21-23, 1976, Paper VI-3.

HANDA, B.K. (1968): Chemical composition of rain water in parts of Northern India: Prel. Studies, Ind. J. Met. & Geophys. 13: 175-180.

HANDA, B.K. (1969): Chemical composition of monsoon rains over Calcutta, Part-I. Tellus, 21: 95-100.

HANDA, B.K. (1969): Chemical composition of monsoon rains over Calcutta. Part 2: Tellus, 21: 101-106.

HANDA, B.K. (1969): Chemical composition of rain water over Calcutta. Ind. J. Met. & Geophys. 20: 150-154.

HANDA, B.K. (1969): Chemical composition of rain water in some parts of northern India. Ind. J. Met. & Geophys., 20: 145-148.

HANDA, B.K. (1970): Chemical composition of pre-monsoon showers over Calcutta. Sci. & Cult. 36: 124-125.

HANDA, B.K. (1971): Chemical composition of monsoon rains over Dankipur, Malda distt. W. Bengal. Ind. J. Met. & Geophys. 22: 603-604.

HANDA, B.K. (1973): Chemical composition of monsoon rain water over Chandigarh in 1971. Ind. Geohydrol. 11: 31-37.

HANDA, B.K. (1973b): Chemical composition of rain water in some parts of India. Symp. on Recent Researches and Applications of Geochemistry. I.A.C.G. & Geochem. Soc. India, Patna. Feb. 1973, p. 4 (Abstract).

HANDA, B.K. (1978): Environmental Pollution: Occurrence of strontium, zinc, copper, lithium and lead in rain water over Lucknow (India): Internat. Symp. on Environmental Agents and their biological effects. Deptt. of Genetics, Osmania University, Hyderabad. Paper Int-S-4-24 to 33.

HANDA, B.K. (1979a): Techniques for Collection and Analysis of rain water samples and presentation of data. Symp. on Chemical Analysis of Geological Materials: Techniques, Application and Interpretation, GSI, Calcutta, Feb. 1979 (Abstract).

HANDA, B.K. (1982): Chemical composition of rain water over Lucknow in 1980. Mausam, 33: 485-488.

HANDA, B.K., KUMAR, A. and GOEL, D.K. (1983): Chemical composition of rain water with special reference to trace elements in Lucknow in 1981. Mausam (in press).

HANDA, B.K. (1977): Modern Methods of Water Analysis. Tech. Man. No. 3; CGWB, 404 p.

JEFFERIES, D.S. and SNYDER, W.R. (1981): Atmospheric deposition of heavy metals in Central Ontaric. Water, Air & Soil Pollution, 15: 127-152.

KUMAR, A., GOEL, D.K. and HANDA, B.K. (1979b): Chemical composition of monsoon rain water over Lucknow with reference to trace elements. Proc. Ind. Sci. Cong. (Abstract).

LANTZY, R.C. and MACKENZIE, F.T. (1979): Atmospheric trace metals: global cycles and assessment of man's impact. Geochim. et Cosmochi. Acta, 43: 511-525.

WIND CHILL INDEX AND THERMAL COMFORT OVER MADHYA PRADESH

S.K. Pradhan
(Meteorological Centre, Bhopal)

Abstract: - Normal Wind Chill Index (WCI) and Thom's 'Discomfort Index' (D.I.) for Madhya Pradesh (M.P.) the Central Indian State, have been calculated. From the distribution of WCI and D.I. over M.P. the following conclusions have been arrived at. January has the highest WCI and hence it is the coldest month in whole of M.P. except the extreme Southeast M.P. where December is the coldest month. Even though the minimum temperatures are lowest in December, the WCI is higher during January. The months of May and June are uncomfortable for M.P. There is serious discomfort over Chambal, Hoshangabad, extreme northern portion of Sagar, Rewa and eastern portion of Raipur divisions during afternoon hours in May. The zone of serious discomfort disappears in the month of June from the State except over Gwalior and Chambal divisions. Very high values of D.I. of the order of 91 or above can occur in extreme cases over some parts of the State which can cause heat strokes.

INTRODUCTION

The existence of man depends upon weather conditions. Food, clothing, shelter, health and happiness of man depend upon meteorological parameters. A study of meteorological conditions which cause discomfort due to cold and heat is of value to local people and to plan the setting up of industries, the estimation of man hours and for tourists from our Country as well as visitors from abroad.

Cold waves affect the health during winter season resulting in deaths of poor people who are unsheltered and unprotected against cold. Siple (1939) and Siple and Passel (1945) evaluated a factor known as Wind Chill Index which represents the cooling power of wind and temperature combination of shaded skin of human beings. It involves the air temperature and wind speed. It is also termed as dry convective cooling power of the atmosphere. Court (1948) has shown that about 75% loss of heat from the body is contributed by the process of dry convective cooling. The Wind Chill Index which gives a measure of man's cold tolerence is very useful.

Siple (1939) expressed a Wind Chill Index (H) which had limitations because it could not be used for temperatures above freezing and for higher wind speeds. However, Siple and Passel (1945) developed a formula

$$H = (\sqrt{100\,v} + 10.45 - v)(33 - t)$$

where H is heat loss in kg calories $M^{-2}H^{-1}$ of the surface.
v is wind speed in m/s.
t is the temperature in °C.

This formula holds good for persons without clothing who are exposed to cold dry winds.

It is well known that temperature and humidity affect the thermal comfort of man. The human system tries to keep the body at a constant temperature of 37° C. The human body is always trying to balance heat production and heat loss so that a constant state is maintained. The process by which this regulation takes place and by which the body adjusts itself to its environment is called homeostasis. The range of ambient temperature when the homeostat mechanism is not stressed is 29° to 31° C for a clothed person. If it is warm the heat loss mechanism is activated and when it is cold the mechanism to conserve heat is brought into action. From experience obtained from airconditioning of rooms, the maximum comfort lies in between 15° to 25° C and Relative Humidity 40 to 70%. However, the interaction of environment on human confort is complex. Various attempts have been made to assess the sensation caused by the environment on the human body.

Thom (1959) has suggested a discomfort index. This discomfort Index D.I. is obtained by a simple linear adjustment applied to the average of the dry bulb and wet bulb readings taken simultaneously.

$$D.I. = 0.4\,(T_d + T_w) + 15$$

where T_d = dry bulb temperature °F
T_w = wet bulb temperature °F

According to Thom, people feel discomfort as the index rises above 70, 50% feel discomfort with index above 75, everybody feels discomfort when the index reaches 79. As the index value crosses 80 discomfort becomes serious. In the Washington metropolitan area, when the index becomes 86 or higher, government regulations permit mass dismissal of employees who are working under these conditions.

In this paper, normal Wind Chill Index (WCI), and Thermal Comfort over Madhya Pradesh, the Central Indian State, have been computed to study the above parameters.

The normal WCI has been computed for the months of November, December, January and February using Siple and Passel's (1945) formula. Thom's Discomfort Index (D.I.) has been computed for the months of January to December.

The data have been taken from the Climatological tables of observatories in India (1931-60).

ANALYSIS

The normal WCI distribution over Madhya Pradesh is presented in Table 1. The WCI for January is presented in Fig. 1. The normal WCI,

TABLE 1: *Normal Wind Chill Index over M.P. (kg Cal $M^{-2}H^{-1}$).*

S.No.	Station	November	December	January	February
1	Gwalior	298	316	345	299
2	Nowgong	232	292	326	302
3	Guna	247	307	337	294
4	Neemuch	245	312	338	298
5	Sagar	277	306	337	285
6	Bhopal	260	297	324	276
7	Hoshangabad	209	249	268	220
8	Indore	286	333	368	326
9	Pachmarhi	298	329	334	311
10	Seoni	260	291	307	262
11	Khandwa	215	255	268	225
12	Satna	235	297	318	283
13	Umaria	251	295	301	269
14	Jabalpur	236	279	290	256
15	Pendra	259	316	331	296
16	Raipur	211	261	260	214
17	Kanker	222	237	231	208
18	Jagdalpur	218	248	246	207
19	Sheopur	257	316	362	294
20	Ratlam	228	282	335	267
21	Chindwara	265	288	306	252
22	Betul	260	287	296	252
23	Ambikapur	319	362	368	322
24	Mandla	235	254	271	239
25	Champa	214	240	254	199
26	Raigarh	195	231	239	186

mean daily minimum temperature and mean monthly wind speed for divisional headquarters viz. Gwalior, Sagar, Bhopal, Hoshangabad, Indore, Jabalpur, Raipur, Ambikapur, Pachmarhi for the months of November, December, January and February are presented in Table 2.

RESULTS

It is seen from Table 1 that Ambikapur (319) has highest WCI followed by Pachmarhi, Gwalior (298) and Indore (286) during November. Ambikapur (362) has highest WCI followed by Indore (333) and Gwalior, Sheopur, Pendra (316) each during December. Ambikapur and Indore (368) have highest WCI followed by Sheopur (362) and Gwalior (345) during January. Indore (326) has the highest WCI followed by Ambikapur (322) and Pachmarhi (311) during February. It is also seen from the table that all the stations over Madhya Pradesh have higher WCI during January with the exception of Raipur, Kanker and Jagdalpur.

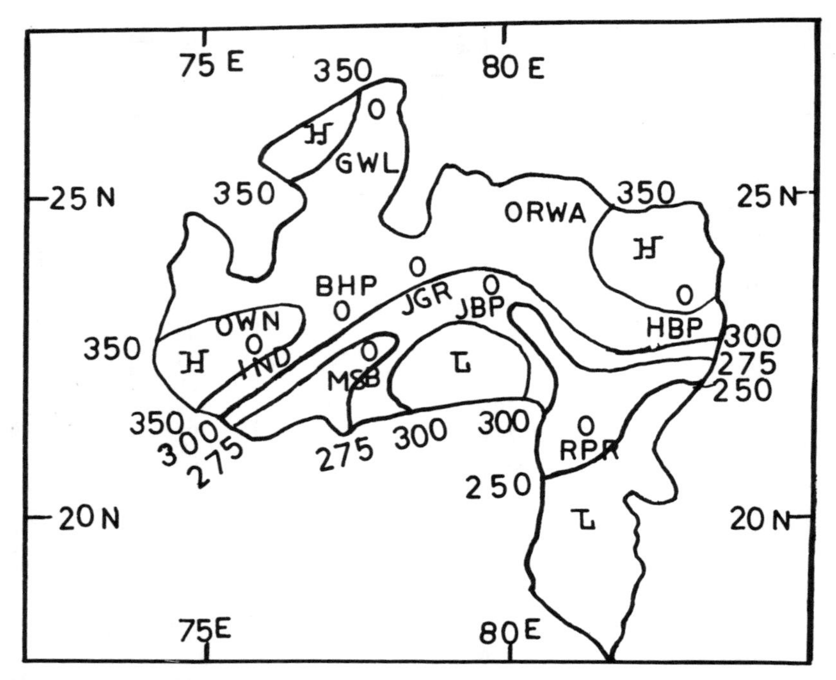

Figure 1: *Normal Wind Chill Index over Madhya Pradesh during January (1931-60). (Iso cooling lines in kg cal.* M^{-2} H^{-1}*).*

It is seen from Fig. 1. that during January the region of higher WCI lies over Bilaspur, Gwalior and Chambal division. A fresh region of high WCI can also be seen over Indore division. The region of lowest WCI is seen over Raipur division. It is seen from Table 2 that over Gwalior, Hoshangabad, Pachmarhi, Jabalpur and Ambikapur the minimum temperatures are lowest in December, but the WCI is higher during January. This suggests that even though the minimum temperatures are lowest in December the chill experienced due to wind is of higher order during January. Bhopal, Indore, Sagar and Ratlam have lowest minimum temperatures and the highest WCI in January. Raipur has lowest minimum temperatures in December and the highest WCI in the same month.

Thermal Comfort:

The computed D.I. values for the various stations in M.P. for January to December are presented in Table 3. To get an idea of extreme possible discomfort over M.P., the D.I. has been computed, taking into account the extreme values recorded as dry bulb, and mean monthly wet bulb air temperatures; the same is presented in Table 4. To get the distribution of D.I. over Madhya Pradesh during the

TABLE 2: *Table showing Wind Chill Index, mean daily minimum temperature and mean monthly wind speed over divisional head quarters of Madhya Pradesh during winter season.*

Station	November			December		
	WC	TN	ff	WC	TN	ff
Gwalior	298	10.5	2.8	316	7.2	2.9
Bhopal	260	13.3	4.3	297	10.6	4.4
Hoshangabad	209	14.5	3.0	249	12.3	3.5
Indore	286	12.1	7.5	333	9.9	7.1
Pachmarhi	298	9.6	2.8	329	7.5	2.7
Jabalpur	236	11.7	2.8	279	9.0	2.7
Raipur	211	16.0	4.1	261	13.2	4.4
Ambikapur	319	10.5	5.8	362	8.0	5.4
Sagar	277	15.5	6.7	306	12.9	6.5
Ratlam	228	14.6	5.9	282	12.3	6.1
	January			February		
Gwalior	345	7.5	3.7	299	10.0	4.4
Bhopal	324	10.4	5.8	276	12.5	6.4
Hoshangabad	268	12.7	4.2	220	14.3	4.2
Indore	368	9.6	9.9	326	11.0	10.8
Pachmarhi	334	8.7	3.3	311	10.4	4.2
Jabalpur	270	9.8	3.4	256	11.4	4.0
Raipur	260	13.5	5.0	214	16.2	6.0
Ambikapur	368	8.6	6.1	322	10.4	6.9
Sagar	337	11.6	6.9	285	13.9	7.1
Ratlam	335	11.0	8.1	267	13.3	7.9

WC = Wind Chill Index in $kg.m.^{-2}\ h^{-1}$
TN = Minimum temperature °C
ff = Wind speed in km/h.

uncomfortable months, i.e. May and June, the D.I. values for afternoon hours are plotted on a map of Madhya Pradesh and the chart was analysed to separate out comfortable and uncomfortable zones. The charts are presented in Fig. 2 and 3 respectively.

The criteria used by Thom for various D.I. values is applicable for colder countries. It is well known that people residing in a place for a sufficiently long time get acclimatized to the normal weather conditions of that place and hence the criteria used by Thom require modification for the Indian climate.

From the experience gained by airconditioning of rooms, the maximum comfort lies between 15° to 25° C and Relative Humidity 40 to 70%. Taking mean condition on this criteria the computed D.I. value for comfort lies between 65 and 75. Hence following classification is suggested:

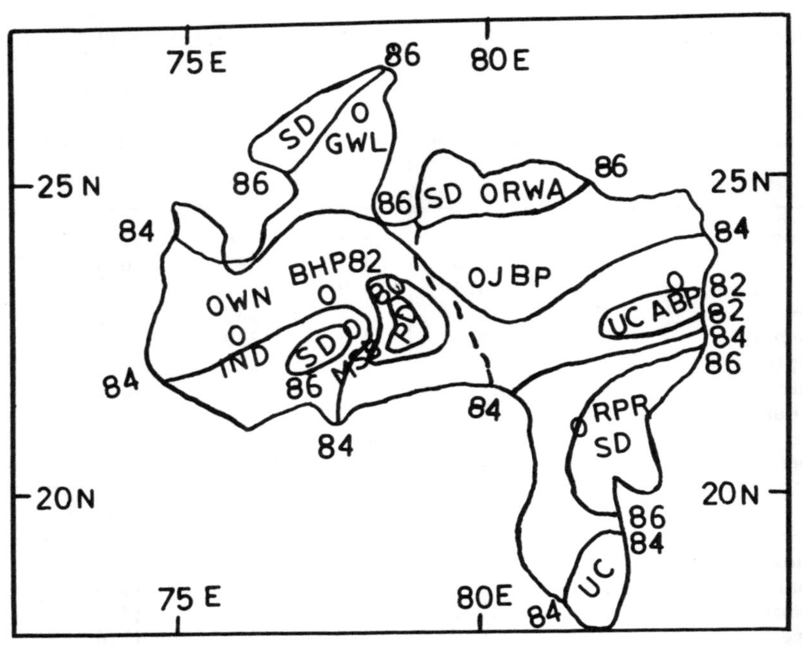

Figure 2: *Mean monthly D.I. over Madhya Pradesh during afternoon hours of May (1931-60).*

D.I. values	less than 65	Uncomfortable
"	66 - 75	Comfortable
"	76 - 79	Partial discomfort
"	80 - 85	Uncomfortable
"	more than 86	Serious discomfort

These modified criteria are used to analyse and study the thermal comfort over Madhya Pradesh.

DISCUSSION

It is seen from Table 3 that the annual variation of D.I. values is from 55 to 87. The lowest range of values are observed during January and February and the highest range during May and June. May and June are uncomfortable as the D.I. values cross 80.

There are stations viz. Raigarh (82), Satna (81), Raipur (81), Kanker (81), Champa (81), Gwalior (81), Nowgong (81), Khandwa (80), Umaria (80) where the D.I, values cross 80 in the morning during the month of May; whereas in the morning of June Gwalior (84), Nowgong (82), Satna (82), Raipur (81), Sheopur (81), Raigarh (81), Umaria (80), Champa (80) where the value of D.I. is more than

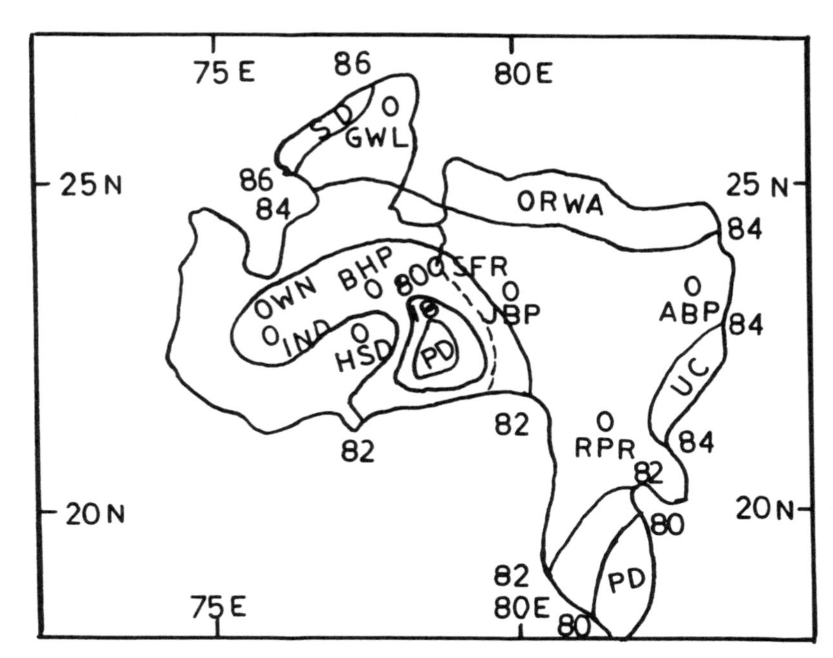

Figure 3: *Mean monthly D.I. over Madhya Pradesh during afternoon hours of June (1931-60).*

80. Where ever the value of D.I. is 80 or more during morning, the discomfort due to heat during day time is more.

The D.I. value crosses 85 during afternoon over Nowgong (87), Champa (87), Raigarh (87), Hoshangabad (86), Satna (86), Raipur (86), Kanker (86), Sheopur (86) in May and over Nowgong (86) and Sheopur (86) in June suggesting serious discomfort over these stations. The D.I. values drop during July suggesting thereby the effect of monsoon on thermal comfort. None of the stations observed D.I. values of the order of 80 during morning hours. However, Gwalior (81), Sheopur (81), Nowgong (80), and Raigarh (80) had D.I. values of the order of 80 or more during afternoon in August and September. This suggests discomfort over these stations during afternoon of August and September.

No station has D.I. values of the order of 80 during October to March suggesting thereby that the State of Madhya Pradesh has comfortable to partial discomfort in these months.

The values of D.I. observed over Pachmarhi suggest that the station has maximum comfort.

Possible extreme D.I. values (Table 4):

It is seen from Table 4 that the highest possible extreme values of 91 are observed over Gwalior, Khandwa, Satna, Raipur, Sheopur,

TABLE 3: *Normal Discomfort Index over Madhya Pradesh*

Station	Month	J	F	M	A	M	J	J	A	S	O	N	D
Gwalior	M	56	60	68	75	81	84	80	79	78	73	64	58
	E	65	70	76	81	85	85	82	81	81	78	71	67
Nowgong	M	55	59	66	75	81	82	79	78	78	72	62	55
	E	67	71	77	83	87	86	81	80	81	78	72	67
Guna	M	56	58	65	73	78	79	77	76	75	70	62	56
	E	67	70	76	80	84	83	79	78	78	76	71	68
Nimuch	M	56	59	66	73	78	79	76	75	75	71	63	58
	E	67	70	75	80	84	83	79	77	78	76	71	68
Sagar	M	59	61	67	73	77	78	76	75	75	71	65	61
	E	66	69	74	79	83	81	78	77	77	74	69	66
Bhopal	M	58	60	67	73	78	78	75	75	75	72	65	60
	E	68	71	75	80	83	81	78	77	77	75	70	68
Hoshangabad	M	61	62	68	74	79	79	77	76	76	72	65	61
	E	70	73	78	83	86	84	79	78	79	78	73	71
Indore	M	58	60	67	73	78	78	75	74	74	72	66	61
	E	69	72	76	80	83	81	78	76	76	75	71	69
Pachmarhi	M	59	61	66	71	75	74	71	70	71	69	63	59
	E	65	67	71	76	79	77	73	72	73	71	67	65
Seoni	M	61	63	68	74	79	76	75	75	75	71	65	60
	E	69	71	76	80	83	80	77	76	76	74	70	68
Khandwa	M	60	62	68	75	80	79	77	76	75	72	65	61
	E	72	75	79	83	85	83	79	78	79	78	74	72
Satna	M	58	61	68	75	81	82	79	78	78	73	64	58
	E	67	71	77	82	86	84	80	79	79	76	71	67
Umaria	M	58	61	69	75	80	80	77	77	77	72	64	58
	E	68	71	76	81	85	83	79	78	78	76	70	67

Champa, Raigarh in May and of the order of 93 over Satna, Champa and Raigarh, followed by D.I. values of the order of 92 over Gwalior, Raipur, Kanker and Sheopur in June. This very high value of D.I. suggests extreme discomfort over these stations which can cause heat strokes.

It is seen from Fig. 2 that the zone of serious discomfort (D.I. Values 86 or more) lies over Chambal, Hoshangabad, extreme northern portion of Sagar, Rewa and eastern portion of Raipur division during afternoon hours in May. It is seen from Fig. 3 that the zone of serious discomfort lies over Chambal division and adjoining Gwalior division, during afternoon hours in June.

The comfortable zone lies around Pachmarhi which is situated over

TABLE 4: *Extreme possible discomfort index over Madhya Pradesh during May and June.*

Station	*May*	*June*
Gwalior	91	92
Nowgong	89	91
Guna	89	91
Nimuch	90	91
Sagar	88	90
Bhopal	88	90
Hoshangabad	90	91
Indore	88	90
Pachmarhi	84	85
Seoni	88	90
Khandwa	91	91
Satna	91	93
Umaria	89	91
Jabalpur	89	91
Pendra	87	89
Raipur	91	92
Kanker	90	92
Jagdalpur	90	90
Sheopur	91	92
Ratlam	89	89
Chhindwara	88	89
Betul	87	88
Ambikapur	87	89
Mandla	89	91
Champa	91	93
Raigarh	91	93

Mahadeo hills during morning hours in May and June. The same zone of comfort around Pachmarhi becomes zone of partial discomfort during afternoon hours. A narrow zone of D.I. values of the order of 78 is observed around Mandla, Pendra and Ambikapur during morning hours in May and June. This can be explained due to the orographic effects of Makal Range situated around 80-81°E and Chotanagpur Ranges situated around 83-84°E. However, this effect of orography over D.I. values is less marked during afternoon hours of May and June.

It is of interest to note that during morning hours the zone of discomfort reduces or remains more or less the same in June as compared to May over the State except over Gwalior and Chambal divisions where the zone increases. This indicates the effect of late onset of monsoon on D.I. The zone of serious discomfort which was seen during afternoon hours of May disappears from Raipur, Bilaspur, and Rewa division, and reduces from Sagar division. The presence of a zone of serious discomfort over Gwalior, Chambal division and extreme northern portion of Sagar division during afternoon hours of June is explained on the basis of late onset of monsoon over this region of Madhya Pradesh.

CONCLUSIONS

(1) Wind Chill Index

The study of WCI over Madhya Pradesh gives a measure of intensity of chill during the winter months. It can be concluded that

i) Since January has the highest WCI, it is the coldest month for whole of Madhya Pradesh except extreme South East Madhya Pradesh where the coldest month is December.

ii) Even though the minimum temperatures are lowest in December over some stations the WCI is higher during January.

iii) The regions of higher WCI, viz. extreme West and extreme East Madhya Pradesh, will require thicker clothing as compared to the other parts of the State to protect from cold.

(2) Thermal Comfort

This study of thermal comfort over Madhya Pradesh lead to the following conclusions:

i) May and June are uncomfortable over Madhya Pradesh. There is serious discomfort over Chambal, Hoshangabad, extreme northern portion of Sagar, Rewa and eastern portion of Raipur division during afternoon hours in May. This zone of serious discomfort disappears or reduces in June from the State except over Gwalior and Chambal divisions.

ii) Very high values of D.I. of the order of 91 or above can occur in extreme cases over some parts of the State which can cause extreme discomfort leading to heat stroke.

iii) A comfortable zone lies around Pachmarhi during morning hours of May and June. The same zone of comfort around Pachmarhi becomes a zone of partial discomfort during afternoon hours.

iv) The effect of orography and onset of monsoon on D.I. values is observed.

ACKNOWLEDGEMENTS

The author is indebted to Dr. A.K. Mukherjee, DDGM, Pune for suggesting the problem. Thanks are due to Shri V.P. Kamble, Regional Director, Nagpur for permitting presentation of the paper and to Shri N.K. Chanchlani for neat typing.

REFERENCES

COURT, A. (1948): Bulletin of American Met. Society, 29.

COMPENDIUM OF METEOROLOGY (1952): Physical Aspects of Human Bioclimatology pp. 1112.

SIPLE, P.A. (1939): Dissertation, Clarke University Library.

SIPLE, P.A. et al. (1945): Proc. Am. Philosophical Society, 89: 1-3.

THOM, E. (1959): Weather Wise, 12: 57-60.

VENKATESHWARAN, S.P. and SWAMINATHAN, M.S. (1967): Indian J. Met. Geophysics, 1: 27-38.

WMO PUBLICATIONS: Technical Note No. 65 and 123.

WIND CHILL OVER HILL STATIONS IN INDIA

K.K. Nathan
(Water Technology Centre, Indian Agricultural Research Institute,
New Delhi - 110012)

Abstract: - The reaction of human beings to temperature and humidity is strongly influenced by the wind speed. With higher wind speed, warm and moist air from near the skin is removed and the person feels cooler. This increases the heat loss of the body to the environment producing "Wind Chill effect". It is also called the dry convective cooling power of the atmosphere. High wind chill effect results in death of many unsheltered and unprotected people in winter. The deaths are not only due to cold waves but also due to many respiratory diseases caused by the cold dry wind. In order to provide first-hand information for the tourists visiting hill stations in India, a "Wind Chill Index" (WCI) has been computed. About 18 stations were selected depending on the availability of normal temperature and wind speed data. The months selected were December, January and February, when the winter is severe and cold waves are prominent. Mean WCI was found to be maximum at Leh (763.94) followed by Dras (746.65) and Srinagar (664.89). The index value in January was higher in almost all the stations ranging from 323.0 to 794.0 as compared to December and February. As the index is meant for the person without clothing, approximate clothing thickness was also computed with available standard nomogram for better protection. The average clothing thickness was found to range between 2 and 14.5 mm among the hill stations in India. The maximum clothing thickness is esperienced at Dras with 14.5 mm followed by Leh (12.0 mm) and Srinagar (9.5 mm).

INTRODUCTION

The response of human beings to temperature and humidity is strongly influenced by the wind speed. Man gains heat when the ambient temperature is higher than skin temperature (say 33° C). However, under conditions in which the air temperature is less than 33° C, man feels colder. At higher wind speeds and at low temperature, convectional heat loss is very much pronounced and it leads to what is known as "Wind Chill Effect". It is sometimes defined as the dry convective cooling power of the atmosphere. This effect causes man to suffer from many respiratory diseases and hypothermic injury within the body by exposing the naked body to the cold dry winds. Prolonged exposure to cold results in the vasoconstriction of the blood vessels which drastically reduces the blood flow in the skin.

In order to study the cooling power of the air, an empirical formula called Wind Chill Index was proposed (Siple and Passel, 1945). This index gives an idea of man's cold tolerance which is based on atmospheric wind speed and low air temperature when the human body is exposed naked to the environment. In this paper an attempt is made to compute and study the WCI for the hill stations in India under mountain environment. About 19 hill stations in India, including one in Nepal were selected (Fig. 1) in order to study the wind chill effect of mountain biometeorology during winter months (December to February). This study will undoubtedly give a first-hand information to the visitors and tourists thronging the hill stations during winter. Since the above formula is for no clothing condition, a minimum clothing is a must to prevent wind chill hazard (Steadman, 1971). A minimum clothing thickness is being computed for various hill stations using the nomogram of Steadman (1971). Apart from the above factors, comfort scales for the various hill stations and normal conduction of the clothing were worked out for the benefit of tourists visiting the hill stations so that they are not caught unaware due to this hazard of wind chill without proper protection.

DATA USED FOR STUDY

In order to study the wind chill effect and its allied parameters like clothing thickness, comfort scale, thermal conductivity etc., mean monthly ambient temperature and wind speed were extracted from the climatological normals published by the India Meteorological Department. These data are available for about 19 hill stations in India including Nepal. These data were collected only for the winter period, namely December to February.

METHOD OF ANALYSIS

Siple (1939) expressed wind chill index as a product of wind speed and temperature. The limitations being that this could not be applied for temperature below zero and for higher wind speeds. However, Siple and Passel (1945) developed an empirical equation overcoming the above limitation. They took many observations of the freezing rate of water sealed in a small plastic cylinder at various temperatures and wind speeds. The wind chill index is given by $H = (\sqrt{100V}+10.45-V)(33-T)$.

H = Heat loss from the surface of a body kg. cals. m^{-2} h^{-1}

V = Atmospheric wind speed, m/s.

T = Ambient temperature °C.

The above expression has been used by the present author in the

Figure 1: *Selected hill stations in India and neighbourhood.*

earlier study of wind chill in Uttar Pradesh region of India (Nathan et al., 1977) where cold waves are prominent. Court (1948) has modified the Siple formula as: H = (10.9 V + 9. 0-V) (33-T).

From the data of wind speed and temperature, clothing thickness were worked out for the hill stations during the winter season using the nomogram of Steadman (1971). (Fig. 2). Having worked out the clothing thickness for each station, thermal conductivity was computed from skin temperature, ambient temperature, clothing thickness and heat flow from the body. Thermal conductivity was also worked out by

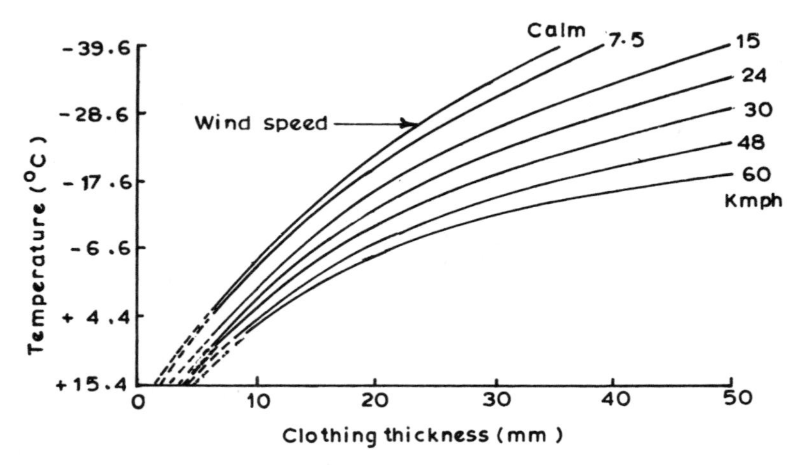

Figure 2: *Thickness of clothing (mm) required to insulate 85% of the body's surface. (R.G. Steadman, Jour. of applied meteorology, 1971).*

replacing skin temperature with normal body temperature. Simple correlation between Index and altitude, Index and temperature and Index with wind speed were tried for the hill stations. Since this paper is for the winter tourists, a relative sensation of comfort scale was worked out based on the wind chill values (Wilson, 1983) for the selected hill stations (Table 2).

DISCUSSION

Table 2 gives an overall information on the various biometeorological factors at different hill stations in India. Using Siple's equation, the average WCI for the winter season was found to be high for Leh (763.94 kg. cals $m^{-2}h^{-1}$) followed by Dras (746.65 kg cals $m^{-2}h^{-1}$) and Srinagar (664.89 kg cals. $m^{-2}h^{-1}$), respectively. The index registered maximum value in all the stations during January.

A simple correlation study between index and ambient temperature among hill stations revealed that both are related significantly ($r = -0.90$) at 5% and 1% levels. Although there existed some relation between index, altitude and wind speed, much significance was not found.

The clothing thickness varies from 2 mm to 14.5 mm among the stations during winter months (Table 1). For Dras Station the mean thickness was 14.5 mm followed by Leh (12.0 mm) and Srinagar (9.5 mm), respectively. During the coldest month January, the clothing thickness requirement ranged from 1.5 mm for Pachmarhi to 16.0 mm for Dras. However, the nature of the fabrics is not taken into

TABLE 1: *Biometeorological parameters at selected hill stations in India.*

Sl. No.	Stations	Mean air temperature (°C)	Mean wind speed ($km.h^{-1}$)	Mean wind chill index (H)kg.Cals. $m^{-2}h^{-1}$	Mean clothing thickness (mm)
1.	Mount Abu	15.6	4.90	359.90	2.00
2.	Cherrapunji	12.7	7.13	453.35	2.50
3.	Coonoor	14.3	5.70	399.82	1.75
4.	Dalhousie	8.5	3.50	473.14	4.50
5.	Darjeeling	7.3	2.50	461.84	4.50
6.	Dharamsala	Data not available			
7.	Dras	13.4	1.33	746.65	14.50
8.	Gulmarg	Data not available			
9.	Kalimpong	12.5	8.70	481.97	3.50
10.	Katmandu	11.1	1.33	353.08	3.00
11.	Kodaikanal	12.77	12.66	519.17	4.00
12.	Leh	-6.2	3.66	763.94	12.00
13.	Mahabaleshwar*	19.3	8.47	322.22	-
14.	Mercara*	19.8	9.80	320.10	-
15.	Mukteshwar	7.0	11.27	648.61	3.00
16.	Mussoorie	7.5	6.97	569.80	5.00
17.	Ootacamund	12.9	3.23	374.46	3.50
18.	Pachmarhi	16.05	3.40	324.55	-
19.	Shillong	10.6	2.93	416.97	4.00
20.	Simla	6.5	3.90	421.88	5.00
21.	Srinagar	-3.4	3.77	664.89	9.50

* Mean air temperature is appreciably high for computing clothing thickness.

consideration as practically all fabrics such as wool, cotton and for give good insulation. Apart from the thickness of clothing, it is important to know how much heat the cloth conducts from a normal individual. In other words, thermal conductivity of the cloth was computed from skin temperature, air temperature, thickness of the cloth and heat flow (Wind Chill Factor). High thermal conductivity values (K) were found for Dras (0.2332 kg. cals. $m^{-1}h^{-1}$ °C) followed by Leh (0.1848 kg. cals. $m^{-1}h^{-1}$ °C) and Srinagar (0.1553 kg.cals. $m^{-1}h^{-1}$ °C), respectively during winter. Thermal conductivity computed by taking body temperature values were found to be less than skin temperature values.

Wind chill equivalent temperature was supposed to be more meaningful than WCI. This is the temperature under light wind condition (2.2 m/s) that has a cooling power to a given combination of actual temperature and wind speed (Steadman, 1971). Only three stations, viz. Dras (-13.6°C), Leh (-7.4°C) and Srinagar (-5.2°C), could be computed for the equivalent temperature as the data was lacking for other stations.

Lastly, on the basis of Siple's wind chill values, a classification was

TABLE 2: *Wind chill index and Sensation scale in hill stations during winter period.*

Average wind chill index (H) about	Sensation of comfort	Hill Station		
		December	January	February
200	Pleasant	Mount Abu, Katmandu, Mahabaleshwar, Mercara, Ootacamund, Pachmarhi	Katmandu, Mahabaleshwar, Ootacamund, Pachmarhi	Mount Abu, Katmandu, Mahabaleshwar, Mercara, Ootacamund, Pachmarhi
400	Cool	Cherrapunji Coonoor, Dalhousie, Darjeeling, Kalimpong, Mussoorie, Shillong	Mount Abu, Cherrapunji Coonoor, Darjeeling, Shillong	Cherrapunji, Coonoor, Dalhousie, Darjeeling Kalimpong, Shillong
600	Very cool	Kodaikanal, Mukteswar, Simla	Dalhousie, Kalimpong, Kodaikanal, Mukteswar, Mussoorie, Simla	Kodaikanal, Mukteswar. Mussoorie, Simla
800	Cold	Dras, Leh, Srinagar	Dras, Leh Srinagar	Dras, Leh, Srinagar

done for sensation of comfort with respect to hill stations in India so that short term winter tourists may be benefitted. The general grading was as "pleasant, cool, very cool and cold" for the 19 hill stations and is depicted in Table 2 using Wilson's (1963) classification. Accordingly, the hill stations Dras, Leh and Srinagar fall in the cold scale of sensation throughout winter period. The stations Katmandu (Nepal), Mahabaleshwar, Mercara, Ootacamund are under pleasant climatic sensation of comfort in winter. Tourists interested in cool to very cool weather may visit stations like Dalhousie, Darjeeling, Shillong, Kalimpong etc. during December to February.

CONCLUSION

Cold stress often induces in man several body reactions in order to maintain the body temperature. A noticeable reaction is shivering, as much of the heat is lost as sensible heat. Under such situation, Siple's formula is a quite useful measure to understand the loss of heat though it has certain limitations. Nevertheless, its usefulness is broadened while computing the clothing thickness as it is mainly based on a temperature and wind speed combination. Stations like Dras, Leh and Srinagar have high values of the index. Thickness of clothing is quite an important issue for tourists engaged in sports and other activities in the hill stations during winter. Clearly the head and the trunk must be covered and protected from cold stress. It is the still air trapped inside the body and cloth that provides good insulation rather than type of cloth. Generally wool, cotton or fur can be used. The station Dras requires maximum thickness about 16.Omm in January and the minimum is for Pachmarhi (1.5 mm). In order to know the cloth that is conducting heat from the body, thermal conductivity is a useful parameter. Similarly equivalent temperature is another important bioclimatic element that is useful for skiers and winter sport lovers at the hill stations. Classification of comfort scale reveals that about 42% of India hill stations falls under pleasant to cool range.

ACKNOWLEDGEMENT

My thanks are due to Dr. A.M. Michael, Project Director, Water Technology Centre, I.A.R.I., New Delhi for his encouragement in writing this paper.

REFERENCES

COURT, A. (1948): Bull. Amer. Meteor. Soc., 29: 487-493.
NATHAN, K.K. and RAO, G.G.S.N. (1977): Vayu Mandal 7: 52-55.
WILSON, O. (1963): Skrifter Norsk Polarinstitutt, 128: 9.
SIPLE, P.A. (1939): Dissertation paper, Clarks University Library.
SIPLE, P.A. and PASSER, C.F. (1945): Proc. Amer. Phil. Soc., 89: 177-199.
STEADMAN, R.G. (1971): Applied Meteor., 10: 674-683.

A NEW CLIMATIC INDEX ON A LINEAR SCALE

V. Kumar
(Defence Science Centre, Ministry of Defence, Defence Research and Development Organisation, Metcalfe House, Delhi-110054)

Abstract: - The present attempt is aimed at integrating major climatic elements for achieving a rational approach in defining a climatic index which should be simple and at the same time enable us to study the climatic features all over the world. A new climatic index, 'Linear Aridity Index', has evolved based on annual averages of three climatic parameters, viz. diurnal range of temperatures, daily mean temperature and precipitation. This index is a measure of aridity of a place, positive for all arid regions and negative for all humid regions, and applicable to high altitudes as well. The climatic scale so derived represents reasonable linearity and is equally sensitive throughout the range. A tentative scheme for the classification of climates has been suggested. The index has been compared against Thornthwaite's moisture index for 32 selected stations in India and neighbourhood (including coastal, non-coastal and high altitude), and a close correlation between the two has been obtained. The climatic pattern in India has also been studied in terms of linear aridity index values. A nomogram has been developed for quick evaluation of the index from the given values of the three parameters used.

INTRODUCTION

Climate, as we are aware, plays an important role in our life and is responsible for determining the living conditions, kind of clothing to wear, production of food, conditions of health and illness etc. The qualitative picture of climates and terrains, based on our past experience, is fairly known but their quantitative description is indeed difficult due to the paucity of meteorological data. In the plains, one is reasonably satisfied if the meteorological stations are placed even 100 km apart because interpolation and extrapolation of data over such distances is more or less permissible, The situation is far more complicated in the case of high altitudes on account of wide variations in the regional microclimates over small areas, depending on altitude and topography.

The important elements which determine the climate are temperature, wind, humidity, precipitation and evapotranspiration. The purpose for which climatic classification is needed may decide about the elements to be incorporated and emphasis is made to make it as quantitative as possible. The problem of classification of climates into various climatic types has been engaging the attention of climatologists all over the world during more than hundred years but little success seems to have been achieved in formulating a general scheme which would meet the requirements of all concerned and at the same time enable us to classify the climates in a quantitative manner.

The early efforts towards classification of climates in a semi-quantitative manner were carried out by Richard Brinsley Hinds during the 19th century. Since then this study is drawing the attention of various workers and the first modern climatic scheme showing the effects of climate on vegetation was described by the German climatologist Koppen (1900) but these approaches have been essentially useful in relation to plant and forest ecology.

THORNTHWAITE'S MOISTURE INDEX

Thornthwaite (1948) introduced a new concept of 'Potential Evapotranspiration', or P.E, which is defined as the total amount of water that would evaporate and transpire if it is always made available for full use. His system of classification of climates based on a critical study of water balance of the soil in relation to certain meteorological parameters is generally regarded as the most rational approach and, perhaps, enjoys the widest popularity.

For the purpose of classification of climatic types, Thornthwaite evolved an empirical formula defining a 'moisture index', I_m, as the dimensionless ratio which is expressed as

$$I_m = \frac{s - 0.6d}{n} \times 100 \qquad \text{...........(1)}$$

where s = surplus moisture in the soil,
d = deficient moisture in the soil, and
n = necessary quantity of moisture for plants (or P.E.)

His simplified expression for the computation of P.E. (or 'n') requires a knowledge of mean monthly temperature and precipitation together with the latitude of the place. According to notations used, positive values of the index represent moist climates and negative values the dry climates. Thornthwaite's scheme for classification of climates in terms of the moisture index has been presented in Table 1.

The concept of potential evapotranspiration and water balance proved useful for the classification of climates but the computational procedure involved in the evaluation of I_m is rather elaborate and somewhat restricts its practical usefulness. The other limitation is the nonlinearity of the climatic scale, being highly compressed for the arid type (I_m: - 60 to - 40) and spread out towards the humid type (I_m: + 20 to + 100). Moreover, no upper limit has been prescribed for the perhumid type where the index value exceeds + 100.

Evidently, the climatic scale with I_m, having a low sensitivity in the arid zone, is not suited for the purpose of grading it in terms of relative degrees of aridity. Based on these considerations, it was

TABLE 1: *Thornthwaite's scheme for classification of climates in terms of moisture index,* I_m.

Climatic type		Moisture index
A	Perhumid	+ 100 and above
B_4	Humid	+ 80 to + 100
B_3	Humid	+ 60 to + 80
B_2	Humid	+ 40 to + 60
B_1	Humid	+ 20 to + 40
C_2	Moist subhumid	0 to + 20
C_1	Dry subhumid	- 20 to 0
D	Semiarid	- 40 to - 20
E	Arid	- 60 to - 40

considered worthwhile to attempt a fresh approach to the problem from a general stand point for evolving a climatic index on a linear scale which should be equally sensitive to all types of climates and easily evaluable from routine meteorological data.

CLIMATIC INDEX ON A LINEAR SCALE

Efforts were initiated in this direction to evolve a climatic index on a linear scale (Majumdar and Sharma, 1964) which should meet the following basic requirements: -

(1) It should reasonably account for the recognised climatic types,

(2) It should, in a fair measure, reflect the combined effect of major climatic elements,

(3) The climatic scale should be equally sensitive throughout the range.

Majumdar and Sharma (1964) considered the concept of diurnal range of temperatures useful, which is simply the difference between daily maximum and minimum temperatures, towards the development of a climatic index. Accordingly, a climatie index, I_{clm}, based on the annual means of diurnal range of temperatures (D) and rainfall (R) was developed which is expressed as

$$I_{clm} = a \log \left(\frac{D/R}{D_o/R_o}\right) \qquad \text{...........(2)}$$

where 'a' is a dimensionless constant and its value depends only on the arbitrary choice of the size of the unit of I_{clm} and D_o/R_o is the value of the ratio D/R when $I_{clm} = 0$, the index being positive for all dry climates and negative for all moist climates.

The above climatic index having linearity on the climatic scale was successfully employed for the sub-classification of arid climates but could not be used with accuracy for high altitudes. Its failure in this respect could be ascribed to non-inclusion of a third important parameter namely, "potential evapotranspiration" or 'n' which is closely related to mean temperature which falls with increasing altitude. Furthermore, the assumption of equal weightage for 'D' and 'R' could not be justified on the basis of available data.

EVOLUTION OF A NEW CLIMATIC INDEX

In order to overcome the defects in the above scheme, the inclusion of the third parameter 'n' was considered and the index so arrived at was termed as 'Linear Aridity Index', I_a. Since a change in I_a should be proportional to the relative changes in D, R and n, and also the index increases with increasing D and n but decreases with increasing R, we may write

$$\delta I_a = a \frac{\delta D}{D} + b \frac{\delta n}{n} - c \frac{\delta R}{R} \qquad \text{...........(3)}$$

where a, b and c are positive constants.

Since both 'n' and 'R' are expressed in cm, the ratio n/R becomes dimensionless but the same cannot be said about the exponent of 'D' which depends on a number of climatic variables. Equation (3) may be put in the form

$$I_a = A \log \frac{D^p.n}{R} \qquad \text{...........(4)}$$

where p being a positive constant. Let the value of the expression $D^p.n/R$ when $I_a = 0$, be denoted by $(D^p.n/R)_o$, so that equation (4) takes the form

$$I_a = A \log \frac{D^p.n/R}{(D^p.n/R)_o} \qquad \text{...........(5)}$$

The linearity of the climatic scale is ensured by the logarithmic scale and the choice of the value of 'A' will determine the size of the unit of I_a.

DEVELOPMENT OF RELATION BETWEEN I_a AND I_m

I_a being an index of aridity, ($-I_a$) may be regarded as a linearised version of Thornthwaite's moisture index, I_m. Equation (5) may be expressed as

$$- I_a/A = \log \frac{R/D^p.n}{(R/D^p.n)_o} \qquad \text{...........(6)}$$

The nonlinear moisture index, I_m, can be obtained by removing the logarithm in the above equation and equation (6) takes the form

$$10^{-I_a/A} = \frac{R/D^p.n}{(R/D^p.n)_o} = q\, I_m + r \qquad \text{...........(7)}$$

By making zero of the I_a scale to coincide with the zero of the I_m scale, we have $r = 1$. Also the limiting value of aridity as per Thornthwaite's scheme is reached with $I_m = -60$ as $R \rightarrow 0$, it follows from equation (7) that $1 - 60q = 0$ or $q = 1/60$, thus equation (7) reduces to

$$10^{-I_a/A} = \frac{R/D^p.n}{(R/D^p.n)_o} = \frac{I_m + 60}{60} \qquad \text{...........(8)}$$

or

$$I_a = A \log \frac{60}{I_m + 60} \qquad \text{...........(9)}$$

In general, we may write

$$I_a = A \log \frac{C}{I_m + C} \qquad \text{..........(10)}$$

where C is a constant having a value close to 60 and can be determined more accurately from an analysis of the data on 'R', 'D' and 'n', together with the computed values of I_m.

EVALUATION OF THE EXPONENT OF 'D'

The computed values of potential evapotranspiration (n, mean annual temperature (t_m), mean annual precipitation (R) and moisture index (I_m) in respect of climates of 32 stations (including coastal, non-coastal and high altitude) in India and neighbourhood were available from a study made by Subrahmanyam (1956). These values together with the computed values of diurnal range of temperatures (D) have been presented in Table 2 and the data utilised for the evaluation of constants of equations (5) and (10).

From equation (8), we find that $(I_m + 60).n/R$ is proportional to D^{-p} or say D^m (replacing $-p$ by m). In Fig. 1, $\log (I_m + 60).n/R$ has been plotted against log D for all the 32 stations. A strong negative correlation is apparent and only four stations (all coastal) shown below the broken line, fall out of the general pattern. Excluding these four points, the mean slope does not differ significantly from -0.5 and we may accept $m = -\frac{1}{2}$ or $p = \frac{1}{2}$. Equation (5), thus takes the form

$$I_a = A \log \frac{n\sqrt{D}}{R} \Big/ \left(\frac{n\sqrt{D}}{R}\right)_o \qquad \text{..........(11)}$$

TABLE 2: *Climatic data of 32 stations in India and neighbourhood.*

S/No	Station	Altitude (m)	t_m* (°C)	D (°C)	R* (cm)	n* (cm)	I_m*	I_a	I_{am}**
1	2	3	4	5	6	7	8	9	10
1.	Agra	169	26.1	15.2	64.4	149.9	- 34.2	+28.2	+25.0
2.	Allahabad	98	25.6	13.2	95.0	147.3	- 18.4	+13.3	+10.9
3.	Calcutta	6	26.1	10.2	162.0	152.0	+ 21.9	- 5.9	- 9.3
4.	Chittagong	27	25.0	8.8	261.0	142.7	+ 84.4	-26.9	-26.4
5.	Darjeeling	2,265	11.7	6.0	303.9	65.5	+364.0	-59.8	-59.0
6.	Delhi	218	25.6	13.5	68.7	146.9	- 31.9	+23.7	+22.5
7.	Gauhati	55	23.9	10.1	170.0	133.6	+ 29.7	-11.8	-12.1
8.	Jodhpur	224	26.7	14.0	34.7	151.4	- 46.3	+46.8	+43.2
9.	Karachi	4	25.0	6.9	19.3	144.0	- 51.5	+50.8	+56.4
10.	Lahore	214	23.9	15.7	51.3	136.2	- 37.4	+37.1	+28.8
11.	Lucknow	113	25.6	13.2	100.0	144.0	- 14.7	+11.8	+ 8.4
12.	Patna	53	22.8	10.4	118.7	148.6	- 6.0	+ 5.1	+ 3.2
13.	Peshawar	354	22.5	13.6	32.1	126.5	- 44.7	+41.0	+40.1
14.	Simla	2,201	13.1	7.2	157.5	70.7	+123.0	-34.4	-33.6
15.	Srinagar	1,586	12.8	13.3	65.7	71.2	+ 6.3	+ 1.0	- 3.0
16.	Veraval	8	25.6	7.6	57.7	148.9	- 36.8	+20.0	+28.1
17.	Bangalore	920	25.0	11.1	87.0	156.4	- 26.6	+12.2	+17.4
18.	Bombay	11	27.1	7.2	192.5	169.4	+ 35.6	-14.6	-14.0
19.	Cuttack	27	27.5	10.4	153.9	169.0	+ 3.5	- 5.3	- 1.7
20.	Cuddapah	130	29.4	11.5	75.5	187.3	- 35.8	+26.6	+26.8
21.	Madras	16	28.3	9.6	125.4	172.3	- 8.4	+ 5.0	+ 4.5
22.	Mangalore	22	27.3	7.1	329.4	173.0	+105.2	-30.8	-30.5
23.	Nagpur	311	24.2	12.2	119.2	153.8	- 6.3	+ 2.6	- 3.3
24.	Pune	558	25.1	13.9	67.9	140.1	- 30.9	+23.4	+21.5
25.	Trivandrum	64	26.4	5.3	163.2	163.6	+ 7.8	-15.6	- 3.7
26.	Ootacumund	2,243	14.2	9.5	139.6	69.6	+100.5	-24.5	-29.6
27.	Port Blair	80	26.3	6.3	313.1	159.9	+103.5	-32.9	-30.1
28.	Rangoon	5	27.4	8.5	261.8	173.5	+ 81.2	-20.9	-25.7
29.	Akyab	9	25.9	7.8	515.3	152.9	+244.6	-45.6	-49.0
30.	Mandalay	77	26.8	11.5	87.1	159.9	- 27.3	+16.0	+18.0
31.	Sukkur	67	26.8	14.4	6.3	152.3	- 57.5	+87.1	+86.5
32.	Colombo	7	26.9	6.0	236.5	168.7	+ 40.5	-24.0	-15.5

* Subrahmanyam (1956)

** I_{am} = linearised version of I_m according to equation (17).

RELATIONSHIP BETWEEN POTENTIAL EVAPOTRANSPIRATION (n) AND MEAN ANNUAL TEMPERATURE (t_m)

The computation of potential evapotranspiration (n) by Thornthwaite's method is rather elaborate and troublesome since it involves the latitude of the place from which a theoretical value of incoming solar

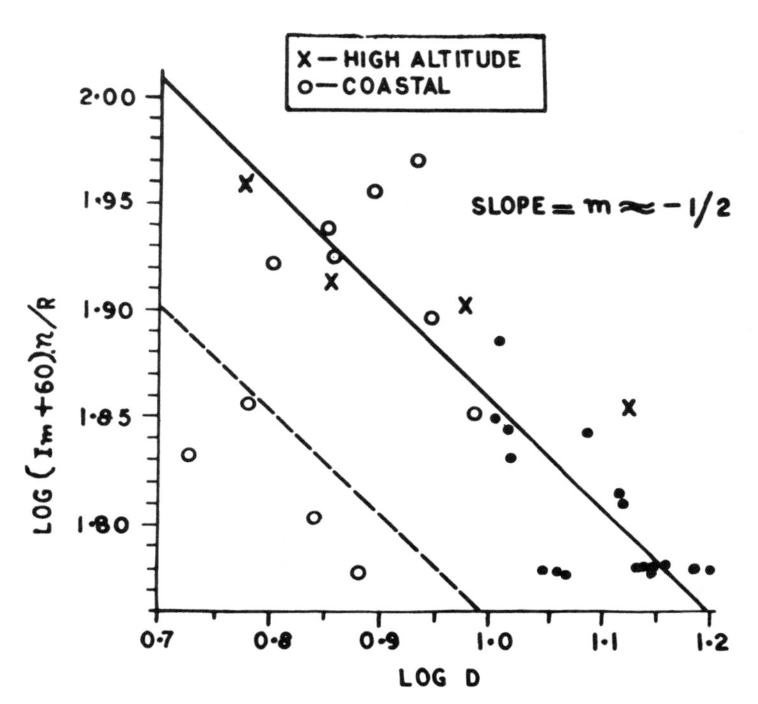

Figure 1: *Evaluation of 'm' in the exponent of 'D'.*

radiation is derived and, therefore, ignores the local factors like cloudiness, turbidity etc. Since the vapour pressure, which determines the rate of evaporation, increases almost exponentially with temperature, it is thought that log n should be linearly related to t_m. In Fig. 2, 'n' has been plotted against 't_m' on a semi-logarithmic scale for all the 32 stations and the expected relationship is quite evident. The regression equation statistically fitted to the points by the method of least squares is

$$n = 31.85 \times 10^{0.0264 t_m} \qquad \text{...........(12)}$$

RELATIONSHIP BETWEEN I_a AND I_m IN TERMS OF R, D AND t_m

With the help of equation (12), we may put equation (11) in the form

$$I_a = A \log \frac{\sqrt{D} \times 10^{0.0264\, t_m}}{k.R}$$

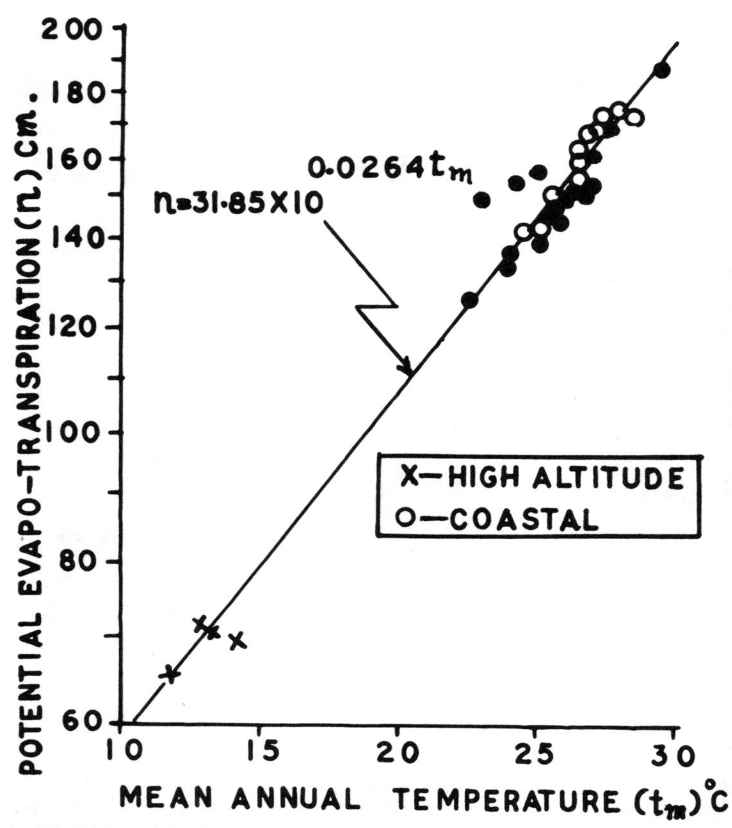

Figure 2: *Relationship between potential evapotranspiration (n) and mean annual temperature (t_m).*

or

$$-I_a/A = \log \frac{k.R.}{\sqrt{D} \times 10^{0.0264\, t_m}} \qquad \text{...........(13)}$$

Also from equation (10), we have

$$-I_a/A = \log \frac{I_m + C}{C} \qquad \text{...........(14)}$$

From equations (13) and (14), it follows that

$$I_m = Ck \,. \frac{R}{\sqrt{D} \times 10^{0.0264\, t_m}} - C \qquad \text{...........(15)}$$

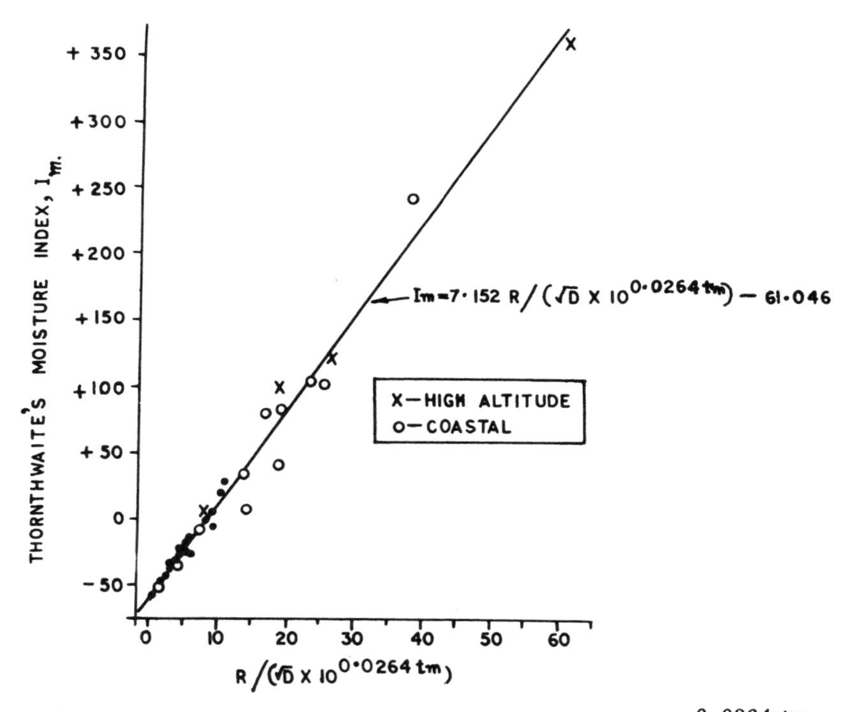

Figure 3: *Linear relationship between 't_m' and $R/(\sqrt{D} X\ 10^{0.0264\ tm})$.*

In Fig. 3, I_m has been plotted against $R/\ \sqrt{D}\ X\ 10^{0.0264\ t_m}$ which yields a straight line and the regression equation obtained is

$$I_m = 7.152\ R/\ (\sqrt{D}\ X\ 10^{0.0264t_m}) - 61.046 \qquad \text{..........(16)}$$

which gives the value of C = 61.046 and equation (10) takes the form

$$I_a = A \log \frac{61.046}{I_m + 61.046} \qquad \text{..........(17)}$$

Since the value of A may be chosen arbitrarily, say A = 70 for convenience, so that I_a may lie roughly between -100 to +100 for the entire range of practical conditions all over the world. Substituting the value for I_m from equation (16) and after some simplifications, equation (17) finally reduces to

$$I_a = 65.19 + 70 \log \frac{\sqrt{D}\ X\ 10^{0.0264t_m}}{R} \qquad \text{..........(18)}$$

The index value is the measure of aridity of a place, being positive for arid regions and negative for humid regions. A nomogram, as

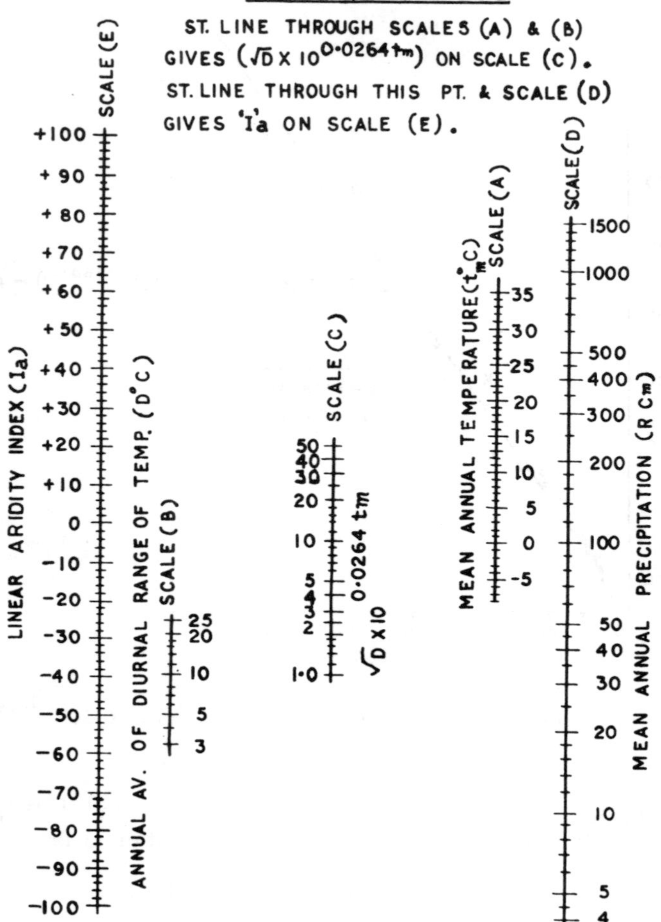

Figure 4: *Nomogram for evaluation of linear aridity index (I_a).*

presented in Fig. 4, has also been developed from which I_a can be quickly evaluated from the given values of D, t_m and R.

COMPARISON BETWEEN I_a AND I_m

With the help of the nomogram in Fig. 4, I_a was computed for all the 32 stations listed in Table 2 (column 9) and compared against reported I_m values as shown in Fig. 5. The closeness of agreement between the

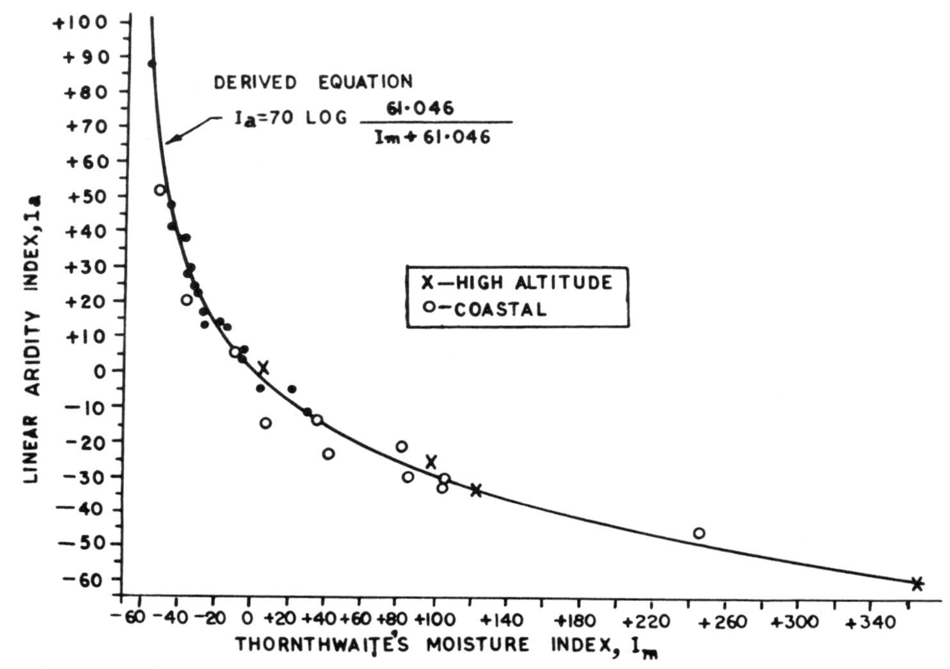

Figure 5: *'I'$_a$ compared against 'I'$_m$ for 32 stations in India and neighbourhood.*

two indices is quite apparent and the curve in the figure represents equation (17) with A = 70.

However, for a quantitative estimation of degree of correlation between I_a and I_m, the linearised version of I_m, designated as I_{am} has been used as given by equation (17). The computed values of I_{am} have been presented in column 10 of Table 2. In Fig. 6, I_a has been compared against I_{am} which reveals a high degree of positive correlation between the two sets of values and the statistically computed coefficient of correlation comes out to be + 0.99.

CLASSIFICATION OF CLIMATES AND SUB-CLASSIFICATION OF ARID AND PERHUMID ZONES

The new climatic index, so derived, has reasonable linearity on the climatic scale and is equally sensitive throughout the range. It has, therefore, been possible to sub-classify arid and perhumid zones and to study the climatic features all over the world in terms of this index. A tentative scheme for classification of climates in terms of linear aridity index has been suggested as presented in Table 3.

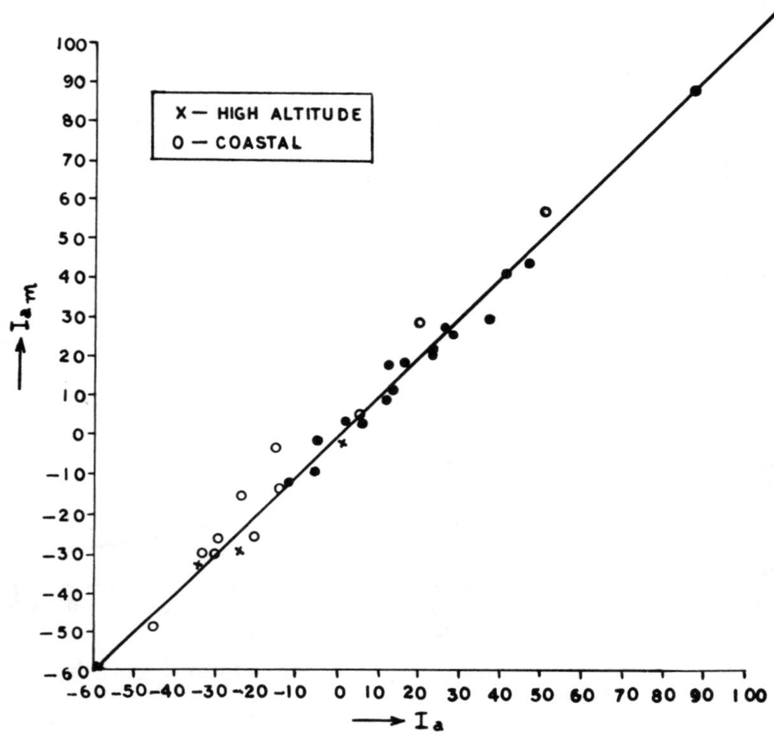

Figure 6: *Comparison between I_{am} and I_a for 32 stations.*

TABLE 3: *Tentative scheme for classification of climates in terms of linear aridity index, I_a.*

Climatic type		Range of I_a
ARID		
	Extreme	$I_a > +70$
	Moderate	$+50 < I_a \leqslant +70$
	Mild	$+30 < I_a \leqslant +50$
SEMIARID		$+10 < I_a \leqslant +30$
SUBHUMID		
	Dry	$0 < I_a \leqslant +10$
	Moist	$-10 < I_a \leqslant 0$
HUMID		$-30 < I_a \leqslant -10$
PERHUMID		
	Mild	$-50 < I_a \leqslant -30$
	Moderate	$-70 < I_a \leqslant -50$
	Extreme	$I_a < -70$

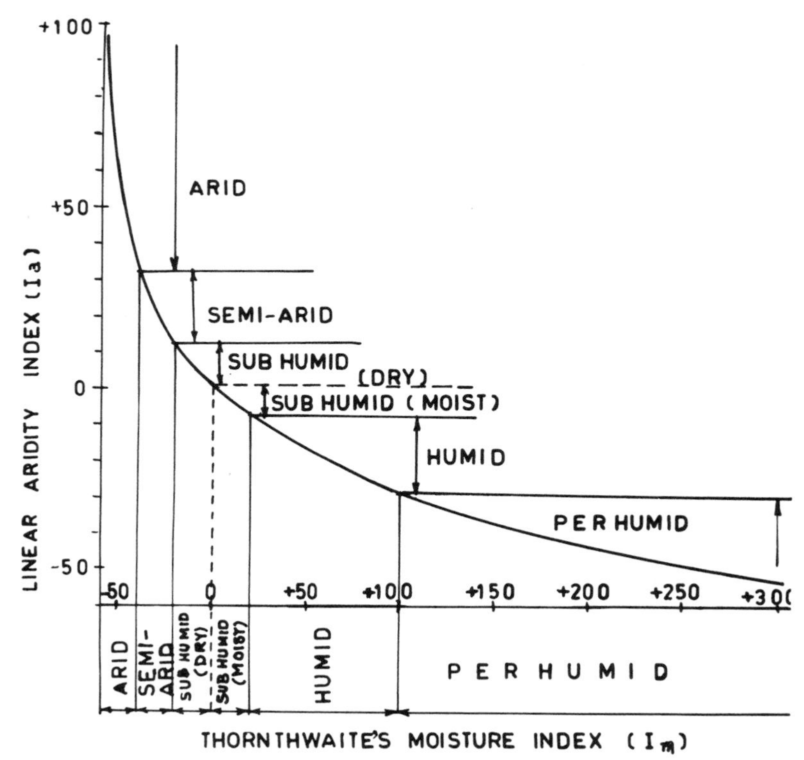

Figure 7: *Broad classification of climates in terms of* I_m *and* I_a.

The curve in Fig. 7 represents the broad classification of climates both along the I_m and I_a scale. It will be observed that the former is highly compressed in the arid zone and gradually spreads out towards the humid and perhumid zones. The I_a scale, on the other hand, is more or less uniformly distributed throughout the zones, each zone covering a range of 20 units. This further supports that the I_a scale is reasonably linear and is equally sensitive for all the climatic zones.

For the purpose of studying climatic pattern in India, 330 stations in India and neighbourhood have been studied by Kumar (1982) in terms of linear aridity index values. The study reveals that most of western Rajasthan is mildly arid with the exception of Jaisalmer (I_a = 61.8), Phalodi (I_a = 57.0), Bikaner (I_a = 50.7), Ganganagar (I_a = 50.4) and Barmer (I_a = 50.2) which are moderately arid. Aridity increases further west and ranges from moderate to extreme in Pakistan, Iran, Arabia and so on. Among the perhumid stations, Cherrapunji and Gnatong top the list with I_a values of -88.5 and -78.2 respectively. Mahabaleshwar (I_a = - 61.4), Darjeeling (I_a = - 53.3) and Gangtok (I_a = - 52.0) are moderately perhumid. Most of Kerala is mildly perhumid.

CONCLUSION

The new climatic index, based on routine meteorological data, represents linearity on the climatic scale and is equally sensitive for all types of climates. It is easily evaluable and is capable of reflecting the combined effect of major climatic elements like humidity, turbidity, wind, solar radiation, cloudiness, nature of terrain etc. in the form of diurnal range of temperatures. Incorporation of the concept of daily mean temperature in the index has made the climatic scale effectively applicable to high altitudes as well. The index can be utilised towards development of agricultural climatology for the purpose of assessment of agricultural potential of unexplored regions and, in particular, at high altitudes. In general, the index serves the purpose of characterising the climatic features all over the world in terms of degree of aridity.

REFERENCES

KOPPEN, W. (1900): Versuch einer Klassification der Klimate, vorzugsweise nach ihren Beziehungen zur Pflanzenwelt. Hettner's Geogr. Zeitschr., vi.

KUMAR, V (1982): 'A biomathematical study of climatic indices'. Ph.D. thesis. Meerut University, Meerut (India).

MAJUMDAR, N.C. and SHARMA, R.N. (1964): A new approach towards classification of climates. Def. Sci. Jour. (India), 14: 161-179.

SUBRAHMANYAM, V.P. (1956): Climatic types of India according to the rational classification of Thornthwaite. Ind. J. Met. Geophys, 7: 253-264.

THORNTHWAITE, C.W. (1948): An approach towards a rational classification of climate. Geog. Rev., 38: 55-94.

PHYSIOLOGICAL ADAPTATIONS TO DESERTS: SWEATING AND HYPOHYDRATION IN MAN

M.K. Yousef
(Department of Biological Sciences, Desert Biology Research Centre, University of Nevada, Las Vegas, Nevada 98154, USA)

Abstract: - The limitations imposed by desert heat on man's performance with special reference to age, sex and race have been under intense investigation in our laboratory. Age, perse, did not significantly reduce the elder's ability to tolerate the combined stress of dry heat and exercise. Caucasian and American Black men had higher sweat rate, lower skin temperature, and heart rate than women working at the same percentage of aerobic capacity. Success of thermoregulation at a work level of 40% of aerobic capacity of American Blacks and Caucasians was equal, but in both races men thermoregulated more successfully than women. The data suggest that thermoregulatory capacity of man under desert conditions differs between sexes and is not influenced significantly by age or race except for differences in aerobic capacity. Physical fitness determined the capacity for working successfully in desert heat and superior performance was related to man's ability to maintain adequate circulatory stability.

INTRODUCTION

One of the most challenging problems facing man in desert enviroments is his ability to cope with heat and aridity. Man utilizes remarkably efficient physiological and behavioral compensatory mechanisms in addition to various technological and sociocultural techniques to insure a successful adaptive capacity to meet the desert climatic challenge. The control of this adaptive capacity operates in a hierarchy at all levels of biological organization, from molecular to organismic, to maintain or restore homeothermy within fairly narrow limits despite diverse and disruptive climatic extremes. For several decades, physiologists have sought to define and reveal the controlling mechanisms of human thermoregulation. This effort has resulted in significant publications wherein voluminous literature is assembled, interpreted and reviewed (Dill 1985; Houdas and Ring 1982; Horvath and Yousef 1981; Tromp and Bouma 1979). Due to limited space, I have confined this paper to the problem of keeping the body cool in hot deserts. Moreover, no attempt has been made to consider the excellent papers resulting from simulation of desert conditions in climatic chambers. This paper reviews only some of the work resulting from studies conducted under desert natural conditions, i.e. field observations.

Homeothermy in Hot Deserts

When faced with increased environmental heat load, men has two primary avenues to prevent an explosive rise in body temperature; sweating and dilation of peripheral blood flow. The former affects the rate at which the body is able to lose heat to its surroundings by evaporation, and the latter enables the body to regulate the flow of metabolic heat from within the body to its skin surface.

Sweating, the only mechanism to maintain homeothermy in a hot environment, results in serious risk of hypohydration due to depletion of body water and salt. In his classic studies on soldiers marching in the Mohave desert, Adolph (1947) found that a failure to voluntarily maintain water balance resulted in considerable hypohydration, even approaching dehydration exhaustion. These levels of water loss led to fatigue, apathy, low morale, unwillingness and inability to undertake strenuous activity and generalized discomfort. What have we learned about sweat gland activity in desert heat? Can hypohydration be minimized or prevented? The answer to these two questions constitute the primary objective of this report.

A. Sweat Gland Activity in Desert Heat

Studies under natural desert conditions have revealed that man, even during moderate work level (walking at a rate of 80-120 m/min), sweats at a level adequate to sustain a relatively constant and regulated body temperature, i.e. about 1.5% of body weight (Yousef, 1980; Yousef et al., 1984; Dill, 1985). Rate of sweating increases to meet the requirement of cooling when thermal stress and work load increase. This finding dispells such popularly held beliefs as, "I sweat very little" or "I sweat profusely." The upper limit of sweat rate while working or exercising in the desert has not been determined. In a recent study, ten young men were able to run on hot days at an average rate of 220 m/min for 1-hr or longer (Smith and Yousef, 1985a). The average sweat rate was 1,350 $g/m^2.hr$ (the range was from 745 to 1740). A question of principle concern was: does the known sweat suppression reported in climatic rooms or under hot humid climates occur in desert walks?

It has been reported that during sustained exposure to heat, sweat rate increases to a maximum, and then decreases (for review: see Cabanac, 1975). However, in our experience sweat "fatigue" or suppression is absent in desert walks (Dill and Yousef, 1976; Dill et al. 1976). Figure 1 records rate of change in total body sweat of 14 subjects, including two women, in a 1-hr desert walk at 100 m/min. The average rate of weight loss rose from 100 g per 7-min lap of walking to about 130 g after three laps and did not change thereafter. In another experiment, 6 men walked continuously at 100 m/min for 5-hr and sweat rate was measured once every hr (Dill et al., 1976).

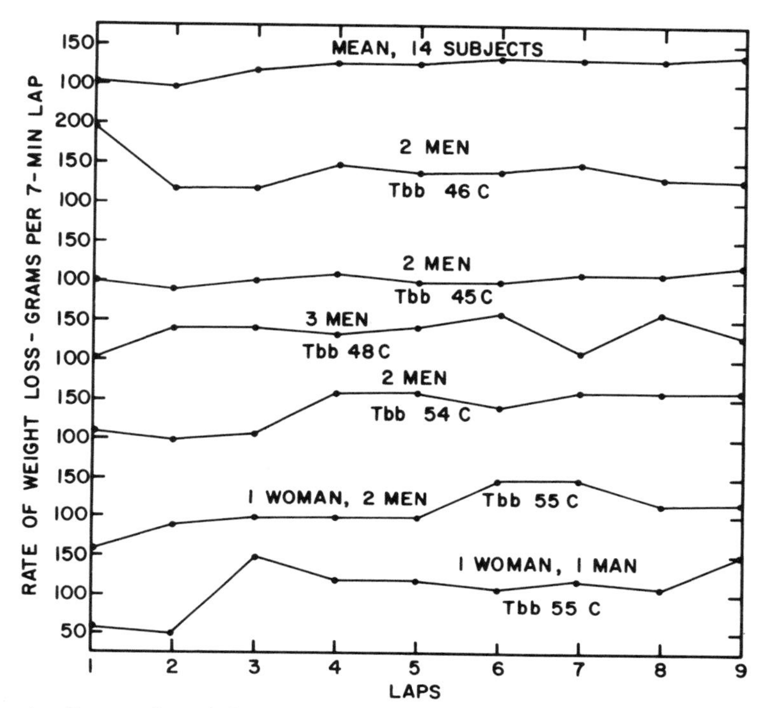

Figure 1: *Rate of weight loss in g per 7 min. for about one-hr (or nine laps). Note the individual differences and that the curve representing the average weight loss for 14 subjects shows no sign of sweat suppression (from Dill and Yousef, 1976).*

The data summarized in Fig. 2 clearly indicate no decrease in sweat rate and a continued rise in rectal temperature and heart rate.

Additional experiments were designed to answer the questions: Does sex, age, or race affect sweat rate? Does drinking water or electrolyte solutions modify sweat rate? and what are the factors affecting sweat composition?

1. Sweat rate and gender:

In the literature, there is considerable disagreement regarding whether or not there are fundamental differences between the sexes in sweat rate. The discrepancy may be attributed to differences in previous state of temperature acclimatization, degree of physical fitness of the subjects and to the type of environmental heat load (i.e. dry vs. humid). In a recent study some of these factors were controlled by using heat acclimatized subjects whose variations in physical fitness

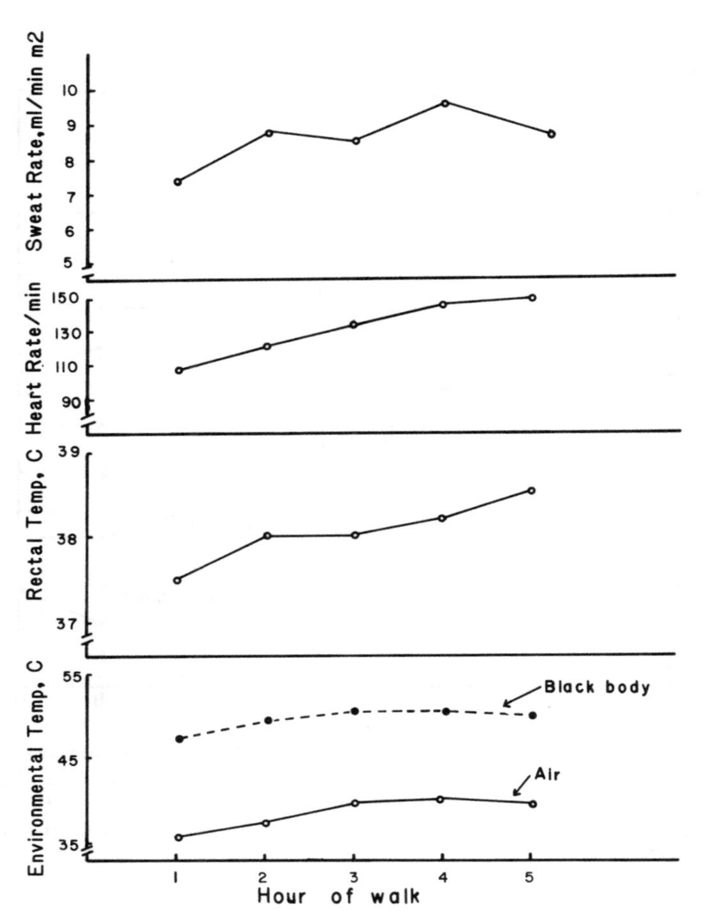

Figure 2: *Environmental temperatures during a 5-hr walk, and the average changes in sweat rate, heart rate and rectal temperature of 6 young men (data from Dill et al., 1976).*

were minimized (Yousef et al., 1984). Sixty women and 57 men ranging in age between 17 and 88 years, walked on hot days at a rate requiring 40% of aerobic capacity of each subject. Regardless of age, women had a significantly lower sweat rate (Fig. 3) and higher skin temperature (Fig. 4), heart rate (Fig. 5) and rectal temperature (Fig. 6). In a previous study, it was concluded that the difference between sexes was more pronounced when rectal temperature increased 1° C or more (Yousef and Dill, 1974).

In a study on effects of dry heat in a climatic chamber, it was concluded that sweat rate for women approaches a limit of about 325 $g/m^2.hr$. In desert walks (100 m/min), Dill (1972) reported a sweat

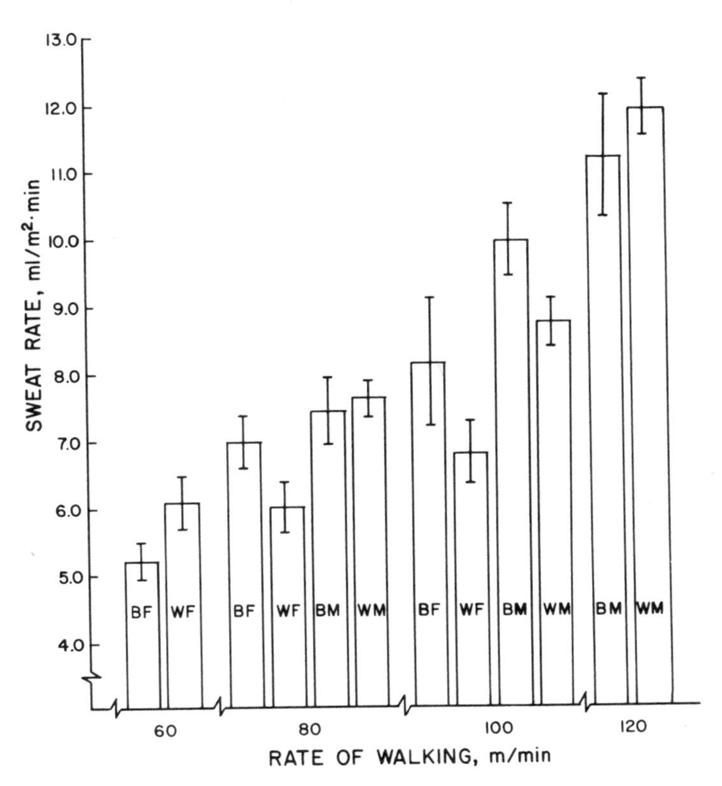

Figure 3: *Average sweat rate of Black women (BF), White women (WF) as compared to Black men (BM) and White men (WM) walking in desert heat at rates ranging from 60 to 120 m/min. The rate of walk for each subject represented 40% of her or his aerobic capacity (from Yousef et al., 1984).*

rate ranging from 302-338 $g/m^2.hr$. Recently, two women (28 - 36 years of age) ran on five different occasions at a rate of 162 to 200 m/min for periods ranging between 50 and 113 min. The average sweat rate for both women was 990 $g/m^2.hr$ with the lowest rate of 840 and the highest rate of 1296 $g/m^2.hr$ (Smith and Yousef, 1985b).

The observed higher skin temperature (Fig. 4) in women than men may be due to decreased sweat rates which in turn may have resulted from adjustments in skin blood flow from the central core to the periphery. In general, our field observations support the hypothesis advanced by Fox et al. (1969) that "women always tend to be less heat-acclimatized than men, even when the two sexes live in the same climate and apparently follow a similar pattern of living. This is because heat acclimatization is essentially a training response, and well-extablished differences between the sexes which are not primarily

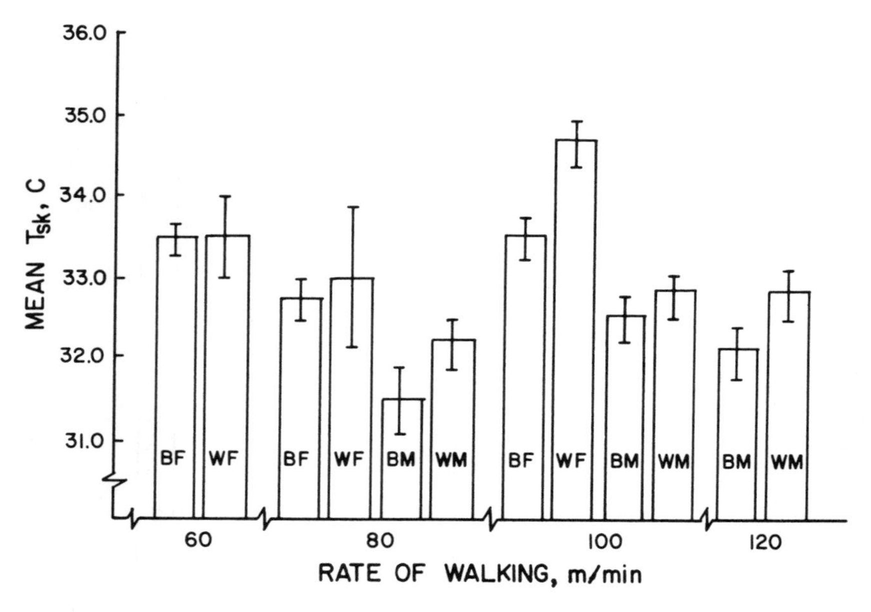

Figure 4: *Mean skin temperature (Tsk) at different rate of walking in men and women. BF and WF represent Black and White women, whereas BM and WM represent Black and White men. (From Yousef et al., 1984).*

thermoregulatory in nature make it easier for women to maintain thermal equilibrium without recourse to sweating."

2. Sweat rate and age:

In desert walks, the capacity for sweating does not decline with age (Dill et al., 1975). The first sign of impending breakdown in desert walks of men ranging in age from 54 to 84 was a high heart rate, not a high body temperature. This suggests that low cardiovascular fitness may be the limiting factor in performance of old men in desert heat. In the literature, there is a discrepancy regarding heat tolerance of elderly people. Shoenfeld and his co-workers (1978) concluded that elderly people should not be exposed to extreme heat for a prolonged period so as to avoid deleterious effects. These observations were made on resting men and women ranging in ages from 46-63 yr who were exposed to 80-90° C and 3-4% relative humidity for a period of 10 min. However, Yousef et al. (1984) demonstrated that elderly subjects as old as 88 yr of age can adequately thermoregulate for a period of 1 hr walking in the desert at air temperatures of 40 to 44° C. The inconsistencies observed in the literature probably are related to various differences in experimental conditions such as the subject's

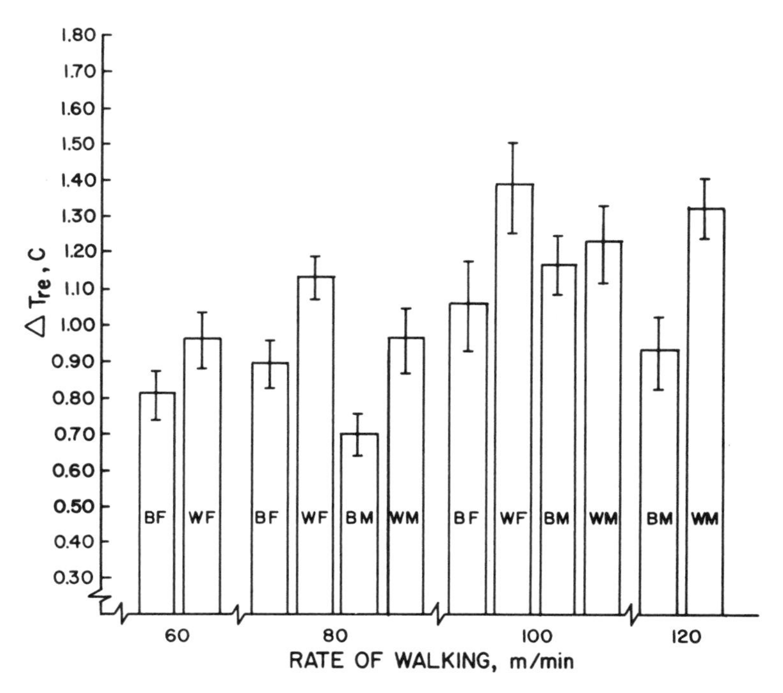

Figure 5: *The increase in rectal temperature (Δ Tre, °C) after walking 1-hr. at different rates in Black and White men and women (notations are the same as in Fig. 4). The pre walk Tre was similar in both sexes and races. (From Yousef et al., 1984).*

physical fitness level, the state of the subject's cardiovascular system, previous state and degree of heat acclimatization; and laboratory versus field experimental conditions. Based on our observations for the past 15 years, elderly men and women who are heat acclimatized, healthy and physically fit can tolerate the combined stresses of heat and exercise at a level requiring 40% of their aerobic capacity, without evidence of deleterious effects.

3. Sweat rate and race:

Of the many studies on man's thermoregulation related to age and gender, only a few are notable in having investigated possible differences between ethnic groups. Furthermore, among these studies from different countries conflicting data on racial differences in thermoregulation have been reported (Baker, 1967; Knipp, 1977; Baker and Weiner, 1966). American Blacks and Whites were studied while working under hot-humid conditions (Robinson et al., 1941; Baker, 1958). These studies showed that blacks were more successful in

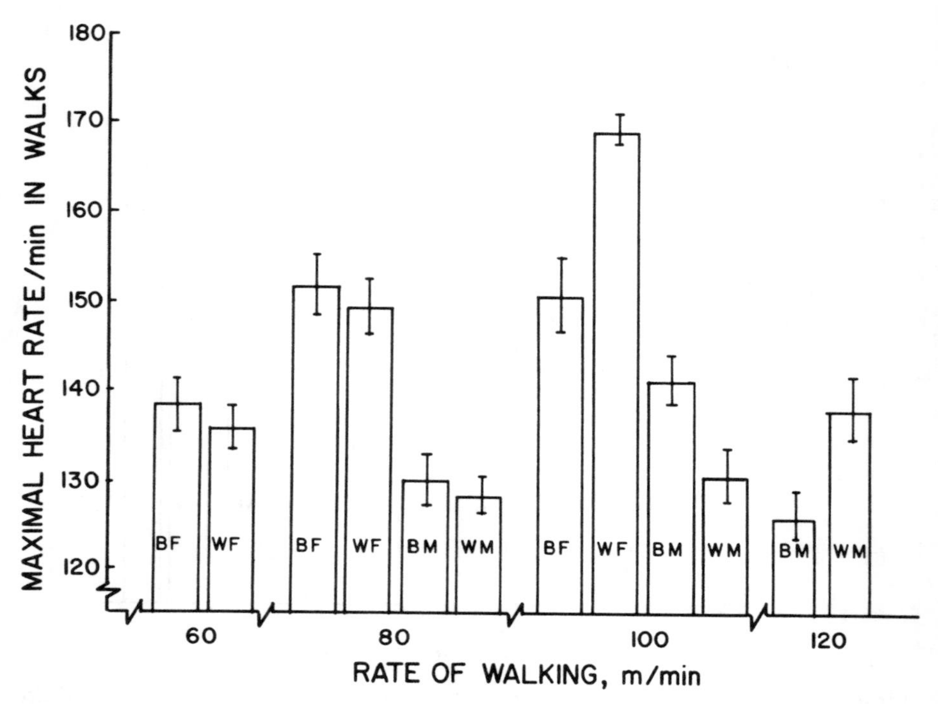

Figure 6: *Maximal heart rate (beats/min) of Black and White men and women walking at different rates in desert heat. Notations are the same as in Fig. 4. The average pre walk heart rate was similar in both sexes and races. (From Yousef et al., 1984).*

thermoregulation than whites. However, under hot-dry conditions both black and white soldiers when clothed and walking or sitting under desert conditions, had about equal heat tolerance (Baker, 1958). When the soldiers were nude and exposed to sun, Whites had higher tolerance than Blacks. In a recent study, when heat acclimatized Black and White men and women of different ages walked at a level requiring 40% of their aerobic capacity, they had similar sweat rate (Fig. 3), rate of rise in rectal temperature (Fig. 6) and heart rate (Yousef et al., 1984). In other words, thermoregulatory capacity was about equal in the young and old in both Blacks and Whites when account is taken of aerobic capacity.

Racial thermoregulatory differences reported among various groups of ethnic people may be attributed to factors other than genetics such as level of previous state of acclimatization, climatic exposure conditions, level of physical fitness, socio-cultural and morphological differences.

4. Sweat composition

Sweat samples may be collected as total body sweat using the washdown methods, or as the hand sweat or arm sweat using hand rubber gloves or arm bags (Yousef and Dill, 1974; Consolazio et al., 1963).

It has been shown by some workers that the various constituents of arm or hand sweat are reasonably representative of the total body sweat. The ratio of chloride concentrations in body sweat to that in hand sweat varied from 0.7 to 0.96 (average 0.75) in men and women walking in the desert (Yousef and Dill, 1974). In all our experiments the composition of sweat at a given sweat rate is found to be an individual characteristic. In general NaCl is the most abundant of all solutes in thermal sweat. Individual values range from 11 to 139; 20 to 118 and 3.7 to 22.8 mEq/l for Na^+, Cl^- and K^+ respectively. Significant variations and wide ranges of all sweat constituents were related to individual idiosyncrasy. For example. one man's sweat may be low in chlorides, high in urea while another's may be high in chlorides and low in urea. Differences in sweat compositions at a given sweat rate were not related to sex, age or race (Dill et al., 1983). In general, sweat chloride tends to increase with sweat rate and age but it decreases with heat acclimatization (Dill et al., 1966). Drinking water or different electrolyte solutions had no significant effect on sweat composition (Dill et al., 1973; Smith and Yousef, 1985a).

5. Sweat rate and drinking:

When man works or exercises in a hot environment, he becomes subjected to various degrees of dehydration even when drinking water is available *ad lib*. With dehydration during exercise in heat, sweat rate was reported to decrease or to remain relatively unchanged. In a study in South Africa, Strydom et al. (1966) assessed the influence of limited water supply on performance of trained, healthy, young men doing an endurance route-march in summer time (environmental conditions during the experimental period were not extreme: dry-bulb temperature did not exceed 32.2° C). They reported no significant difference in total amount of sweat produced in the water-restricted group and the water *ad lib* drinking group. In two-hr walks at a rate of 100 m/min in desert heat, (dry bulb temperature ranging between 37.3 to 47.3 °C), both men and women sweat at the same rate with water and salt replenishment as without drinking water or salt (Dill et al., 1973). Similar data were reported on young men running in desert heat at an average rate of 220 m/min and with a sweat rate as high as 23 g/m^2.min, i.e. about 3.6% of body fluid deficit per hour (Smith and Yousef, 1985a). In general, sweat rate under desert conditions was found to be independent of drinking water or electrolyte solutions of salinities up to 0.9% NaCl.

B. Hypohydration: Can it be prevented?

The delay in complete rehydration following water loss either by sweating or by water deprivation is termed voluntary dehydration

(Adolph, 1947), progressive dehydration (Hertzman and Ferguson, 1960) and involuntary dehydration or hypohydration since man does not reduce his rate of fluid intake "voluntarily" (Greenleaf et al., 1983). Environmental physiologists disagree on the mechanism of hypohydration and on the possibility of preventing this condition.

In 1933, Dill et al. published their pioneering experiment on dehydration and rehydration of a man and a dog who undertook an all-day walk in the desert. At the end of 5.8 hr (about 26 km) the dog's feet were blistered so he stopped while man finished an additional 70 min (a total walk of 32 km). Man finished his walk with a total body weight loss of 3.1 kg, although he drank 6.1 kg water throughout the walk. The sum, 9.2 kg or 13% of body weight was almost entirely sweat. The dog lost 0.1 kg and drank 2.45 kg, or 13% of his body weight. Dill et al. (1933) suggested that a possible explanation of the failure of man to maintain his body weight and the success of the dog to maintain his water balance lies in the loss of salt in sweat. The dog does not sweat; evaporation of water from his air-way tongue entails no loss of salt. Therefore, both man and dog, drank enough to restore osmotic pressure.

The role of salt intake on maintenance of body weight was studied in desert walks by Adolph and his co-workers (1947). Two groups of five men each on the same regime except one group drank water *ad lib* and the second group received enough salt tablets hourly to balance estimated losses in sweat in addition to drinking water *ad lib*. Body weight and fluid intake were measured hourly from 9 am to 4 pm. Due to an arithmetic error in Table 16-3 (page 262 in Adolph's book), it was questioned that body weight can be maintained by taking salt, as well as, water. However, if one corrects the arithmetic error as shown in Table 1, the data indicate that the men receiving salt, had a body weight gain of 110 g and excreted less urine than the other men who drank only water and lost an average of 600 g. The question of whether or not saltiness of the sweat control the amount of water required to satisfy thirst remained open until 1976 (Dill et al., 1976). Seven young and fit men undertook to walk in the desert 30 km in 5 hours with cool tap water to drink each hr (the environmental conditions of this walk are show in Fig. 2). When the concentration of chlorides in body sweat was plotted against the volume of body water used (Fig. 7), it became apparent that the amount of water drunk was less, the saltier the sweat. In other words, water intake was adjusted to the level required for maintaining a constant osmotic pressure.

Can hypohydration be prevented in a desert environment? A series of walks (100 m/min) were undertaken either with no water or with the subjects requested to drink periodically (once every 7 or 8 min) enough water containing an amount of salt to equate their losses of water and salt in sweat (Dill et al., 1973). The average change in body weight with water and salt replenished was -0.12 kg; those without water to drink lost 2.03 kg. The data suggest that subjects walking at a rate of 100 m/min for 2 hr in desert heat and instructed to drink periodically an amount of salt solution that balances salt and water loses in sweat, can prevent or significantly reduce hypohydration by maintaining a relatively constant body weight. However, in a study to reveal the effects of different levels of salt replenishment on hypohydration during heavy exercise (running at

TABLE 1: *Adolph's data (corrected)* on effect of salt intake on voluntary dehydration and urinary flow in a 7-hr walk in desert heat.*

No. of Subjects	Salt intake, g	Total sweat loss, g	Total fluid intake, g	Body Wt change, g	Urine volume, ml
5	0	4,840	4,240	-600	306
5	4.7	4,660	4,770	+110	204

* The original data is in Table 16-3, p. 262 (Adolph, 1947).

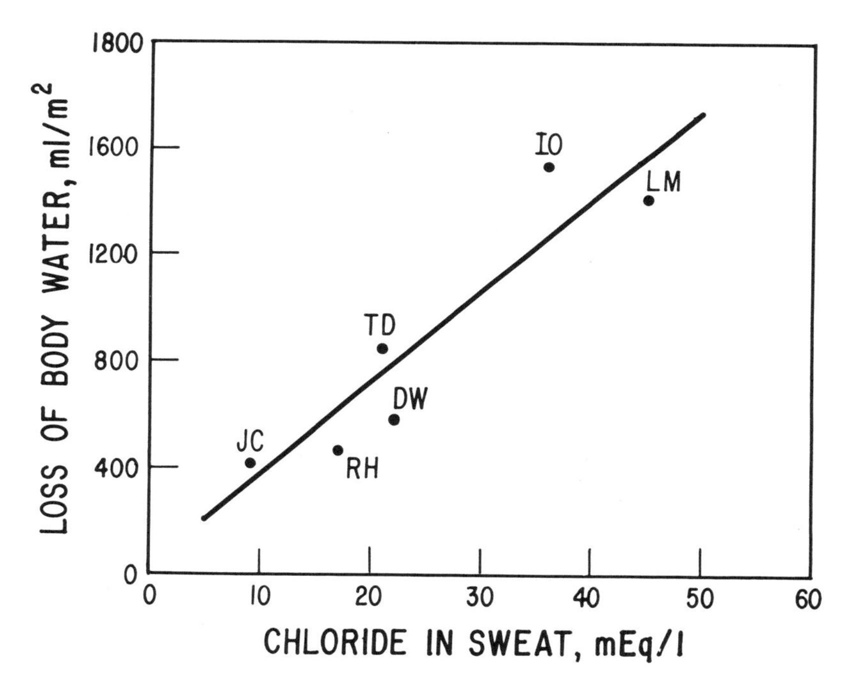

Figure 7: *Relationship of chloride concentration in body sweat of each subject to his loss of body water expressed in ml/m². (From Dill et al., 1976).*

13.5 km/hr) in desert heat, seven young men who drank *ad lib* were unable to replace more than 42% of their body fluids (Smith and Yousef, 1985a). The disagreement between these two studies is perhaps related to differences in experimental conditions, i.e., drinking every 7 or 8 min vs. *ad lib*; walking at 100 m/min vs. 220 m/min; average sweat rate of 9 g/m^2.min.vs. 23 g/m^2 in. It seems reasonable to conclude that hypohydration in desert heat may be prevented with replenishment of salt and water, only if the rate of sweat is at a moderate rate (less than 110 g/m².min or about 1.5% of body weight per hour). Partial maintenance of body fluids (less than 50% replacement) may be accomplished during heavy workload when sweat loss equals 3.5 to 4.0% of body weight per hr.

CONCLUSIONS

Maintaining homeothermy in desert heat depends entirely on evaporative cooling of sweat. Under desert natural climatic conditions, sweat suppression or "fatigue" is not evident even when young individuals walk at 100 m/min for 5-hr. Total body sweat rate is not significantly affected by drinking water or electrolyte solutions, nor by age or race. Each individual sweats sufficiently to prevent an explosive rise in body temperature. Men had higher sweat rates than women working at the same percentage of aerobic capacity. The volume of water drawn from body reserves seems to be closely correlated with concentration of electrolytes in body sweat, and the volume of drinking water that satisfies thirst maintains osmotic pressure. Prevention of hypohydration during desert walks may be achieved if body water loss does not exceed 1.5% per hr by instructing the individual to drink periodically an amount of salt solution that balances his or her salt and water losses in sweat. If body water loss exceeds 3%, water and/or salt replenishments may reduce hypohydration by only no more than 50%.

In general, thermoregulatory capacity of humans under desert conditions differs between sexes and is not influenced significantly by age or race except for differences in aerobic capacity.

REFERENCES

ADOLPH, E.F. and Associates (1947): Physiology of Man in the Desert. Interscience, New York.

BAKER; P.T. (1958): Racial differences in heat tolerance. *Am. J. Phys. Antrop.*, 16: 287-305.

BAKER, P.T. (1967): The biological race concept as a research tool. *Am. J. Phys. Antrop.*, 27: 21-25.

BAKER, P.T. and WEINER, J.S. (1966): The Biology of Human Adaptability. Clareson Press, Oxford.

CABANAC, M. (1975): Thermoregulation. *Ann. Rev. Physiol.*, 37: 415-439.

CONSOLAZIO, C.F., MATOUSH, L.O., NELSON, R.A., HARDING, R.S. and CANHAM, J.E. (1963): Excretion of sodium, potassium, magnesium, and iron in human sweat and the relation of each to balance and requirement. *J. Nutr.*, 79: 407-415.

DILL, D.B. (1972): Desert Sweat Rate. In: S. Ito, K. Ogata and H. Yoshimura (eds.), *Advances in Climatic Physiology*, Igaku Shoin Ltd, Tokyo, 134-143.

DILL, D.B. (1985): The Hot Life of Man and Beast, C.C. Thomas, Springfield, IL.
DILL, D.B., BOCK, A.V. and EDWARDS, H.T. (1933): Mechanisms for dissipating heat in man and dog. *Am. J. Physiol.*, 104:36-43.
DILL, D.B., HALL, F.G. and VANBEAUMONT, W. (1966): Sweat chloride concentration: sweat rate, metabolic rate, skin temperature and age. *J. Appl. Physiol.*, 21: 99-106.
DILL, D.B., KASH, F.W., YOUSEF, M.K., SCHOLT, L.F. and WOLFENBARGER, D.L. (1975): Cardiovascular responses and temperature regulation in relation to age. *Aust. J. Sports Med.*, 7: 99-106.
DILL, D.B., SCHOLT, L.F. and ODDRSHEDE; I. (1976): Physiological adjustments of young men to five-hour walks. *J. Appl. Physiol.*, 40: 236-242.
DILL, D.B. and YOUSEF, M.K. (1976): Dehydration, rehydration, and exercise: dry and humid heat. In: S.M. Horvath and R.C. Jensen (eds.), *Standards for Occupational Exposures to Hot Environments.* USDHEW Pub. No. (NIOSH) 76100: p. 1-6.
DILL, D.B., YOUSEF, M.K., GOLDMAN, A., HILLYARD, S.D. and DAVIS, T.P. (1983): Volume and composition of hand sweat of white and black men and women in desert walks. *Am. J. Phys. Anthrop.*, 61: 67-73.
DILL, D.B., YOUSEF, M.K. and NELSON, J.D. (1973): Responses of men and women to two-hour walks in desert heat. *J. Appl. Physiol.*, 35: 231-235.
FOX, R.H., LOFSTEDT, B.E., WOODWARD, P.M., ERIKSOON, E. and WERKSTROM, B. (1969): Comparison of thermoregulatory function in men and women. *J. Appl. Physiol.*, 26: 444-453.
GREENLEAF, J.E., BROCK, P.J., KEIL, L.C. and MORSE, J.T. (1983): Drinking and water balance during exercise and heat acclimation. *J. Appl. Physiol: Respirat. Environ. Exercise Physiol.*, 54: 414-419.
HERZMAN, A.B., and FERGUSON, I.D. (1966): Failure in temperature regulation during progressive dehydration. *U.S. Armed Forces Med. J.* 11: 542-560.
HORVATH, S.M. and YOUSEF, M.K. (1981): Environmental Physiology: Aging, Heat and Altitude. Elsevier/North Holland, New York.
HOUDAS, Y. and RING, E.F.J. (1982): Human Body Temperature. Its Measurement and Regulation. Plenum Press, New York.
KNIP, A.S. (1977): Ethnic studies on sweat gland counts. In: J.W. Weiner (ed.), *Physiological Variations and its Genetic Basis.* Taylor and Francis, Ltd., London. P. 113-123.
ROBINSON, S., DILL, D.B., WILSON, J.W. and NIELSEN, M. (1941): Adaptations of white men and Negroes to prolonged work in humid heat. *Am. J. Tropical Med.*, 21: 261-287.
SHOENFELD, Y., UDASSIN, R., SHAPIRO, Y., OHRI, A. and SOHAR, E. (1978): Age and sex difference in response to short exposure to extreme dry heat. *J. Appl. Physiol.: Respir. Environ. Exercise Physiol.*, 44: 1-4.
SMITH, M.S. and YOUSEF, M.K. (1985a): Evaporative cooling of man running in desert heat. In: *Seventh Conference on Biometeorology and Aerobiology*, Am. Meteor. Soc., Boston, Mass. (in press).
SMITH, M.S. and YOUSEF, M.K. (1985b): Sweat rate and composition in men and women running in desert heat. *Fed. Proc.*, 44: 1562.
STRYDOM, N.B., WYNDHAM, C.H., VAN GRANN, C.H. HOLDWORTH, L.D. and MORRISON, J.F. (1966): The influence of water restriction on performance of men during a prolonged march. *South Afr. Med. J.*, 40: 539-544.
TROMP, S.W. and BOUMA, J.J. (1979): Biometeorological Survey, Vol. I, 1973-1978, Part A, Human Biometeorology, Heyden, London.
YOUSEF, M.K. (1980): Responses of men and women to desert heat. *Int. J. Biometeor.* (Supplement). 24: 29-41.
YOUSEF, M.K. and DILL, D.B. (1974): Sweat rate and concentration of chloride in hand and body sweat in desert walks: male and female. *J. Appl. Physiol.*, 36: 82-85.

YOUSEF, M.K., DILL, D.B., VITEZ, T.S., HILLYARD, S.D. and GOLDMAN, A.S. (1984): Thermoregulatory responses to desert heat: age, race and sex. *J. Gerontology*, 394.

PHYSIOLOGICAL RESPONSES DURING PROLONGED WORK IN HOT DRY AND HOT HUMID ENVIRONMENTS

J. Sen Gupta
(Defence Institute of Physiology and Allied Sciences
Delhi Cantt - 110010)

Abstract: - A progressive increase in climatic stresses during work will result in physiological strains on the body leading to incapacitation. Studies were conducted on prolonged continuous work of varying rates at comfortable, hot dry and hot humid environments. Results indicated that above a critical air temperature, body temperature rises to a new higher steady level but in extreme hot conditions, deep body temperature rises along with mean skin temperature leading to heat exhaustion. Thus, the problem of assessment of the individual capacity for sustained physical effort is further complicated and no suitable physiological index is available for the prediction of work tolerance in heat. It is, therefore, proposed to combine a number of functional aspects into a single physiological index to predict endurance work capacity in hot dry and hot humid environments.

INTRODUCTION

Man has to work for his survival and existence in any environment whether it is arctic cold or tropical heat. It is well recognised that work output of farm labour or industrial labour is lower in summer months than in the winter months in India. Besides the work output, climate of a country has general effects upon the culture of the population including working habits, modification of clothing and housing. Extremes of heat (dry or humid) affect work capacity and limit prolonged continuous work due to (i) accumulation of heat in the body (ii) thermoregulatory disturbances leading to body dehydration, and (iii) loss of electrolytes.

ASSESSMENT OF ENDURANCE WORK CAPACITY IN MAN

The problem of assessment of an individual's capacity for prolonged work from physiological measurements is quite complex and is further complicated when work is performed in hot environments. The various indices so far developed for the purpose are of only limited use.

Muller (1950) proposed the work pulse index (LPI) as a measure of occupational work capacity which he has defined as the highest permissible work capacity. On the basis of his LPI, Muller (1962) recommended an average work load of 20% of maximum 0_2 uptake capacity for 8 hr work schedule for healthy people. An interesting finding is that while Vo_2 max declines with increasing age, occupational work capacity remains unaltered upto 70 years of age. At the same time, Lehmann et al. (1950) assumed 2500 KCal to be available during a working day of 8 hr. These figures, however, do not take into account the large individual variations. Bink (1962) and Bonjer (1962, 1968) plotted a diagram showing the relationship between allowable energy expenditure and the working time to indicate the energy expenditure that can be maintained throughout the working time by an individual with a given aerobic capacity. The linear relationship between physical working capacity and the logarithm of the working time is expressed by the formula

$$\mathring{A} = \frac{\text{Log } 5700 - \text{log.t}}{3.1} \times a \text{ KCal/min}$$

where $\mathring{A}$ = Physical working capacity in KCal/min
t = Working time in min
a = aerobic capacity in KCal/min

According to this formula, the endurance time in any physical effort can be predicted for an individual from the knowledge of his aerobic capacity and the work load. Actual computation with this formula indicates that 36% of maximum aerobic capacity can be endured for an 8 hr period, 50% for 2 hr 50 min. Similar conclusions have been made by Michael et al. (1961) who observed that an 8 hr work-rest schedule could be completed without undue fatigue when the energy cost did not exceed 35% of maximum and the heart rate remained below 120 beats/min. These workers further concluded that a heart rate of 140 beats/min could not be maintained for more than 4 hr or a heart rate of 160 beats/min for more than 2 hr without extreme fatigue. Astrand (1967) reported that, when the work period is extended in 1 hr or more, the oxygen uptake, heart rate and cardiac output are maintained at a steady state provided the exercise 0_2 uptake is within 50% of the maximum. Costill (1972), on the other hand, pointed out that champion marathon runners could sustain 84% of their $\dot{V}o_2$max for more than 2 hr. Pruett (1970), on the basis of energy supply, reported that 20% of maximum work can be sustained for 6 hr with 45 min work followed by 15 min rest. Thus, there appears to be little agreement among various workers regarding the work load recommended for different work-rest schedules for 8 hr work, although all of them have expressed the work load as a fraction of $\dot{V}o_2$max. We have reported earlier (Sen Gupta et al., 1972) that exercise dyspnoea gives a good indication of a person's endurance capacity for different work rates. In another study, (Sen Gupta et al., 1974) we have shown that an index based on combined cardio-pulmonary strains imposed by work yielded a highly significant relationship with the endurance time in work. (Fig. 1).

The proposed method was considered to be very useful for the prediction of work tolerance, because of its simplicity and reasonable

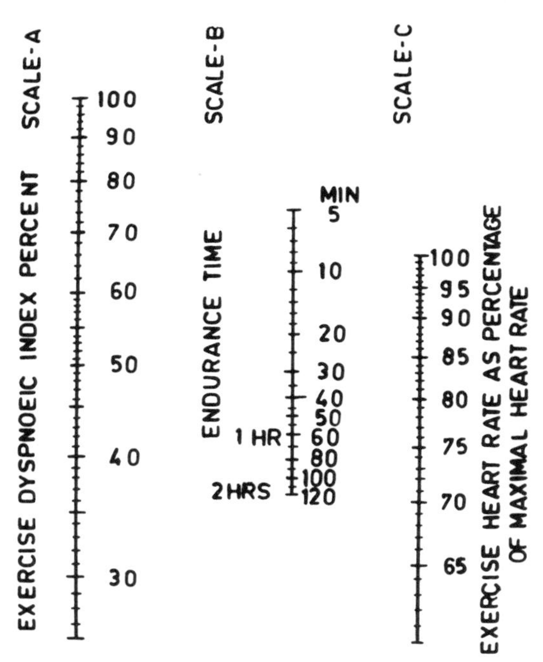

Figure 1: *Nomogram for prediction of endurance time from exercise dyspnoeic index and exercise heart rate as percentage of maximal heart rate.*

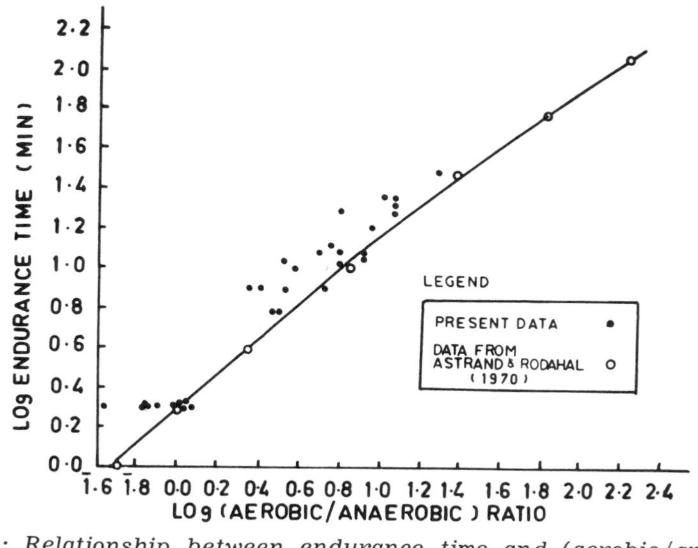

Figure 2: *Relationship between endurance time and (aerobic/anaerobic) ratio of oxygen supply.*

degree of reliability. Another alternate approach (Sen Gupta et al., 1974), based on aerobic and anaerobic fractions of 0_2 supply during work, has been found to be quite satisfactory as a single index as the equation ($T = Au, ^{K_1} u_2 - K_2$) reported by Astrand and Rodahl (1970) fits the data of an athlete perfectly over the entire range of duration (1-120 mins) as well as 2-70 mins duration in endurance efforts by average Indians (Fig. 2).

PHYSIOLOGICAL AND METABOLIC RESPONSES DURING WORK IN HOT ENVIRONMENT

During muscular work, the body temperature is regulated with a thermostat set at a higher level and this level is directly related to rate of work and the level is unaffected by air temperature over a wide range. This has been reported by Wyndham et al. (1953) and Lind et al. (1963), which has been further confirmed by us. It has been observed that the relationship between rectal temperature and metabolic rate holds true only upto certain critical air temperatures which are different for different metabolic rates. It has been further observed, that during work in high ambient heat, rectal temperature continues to rise leading to incapacitation or termination of work. The effects of heat on the TR/Vo_2 max relationship are different in acclimatized and unacclimatized subjects. It has been further observed that the TR/Vo_2 max relationship during exercise in heat is of some practical importance in selecting men for work in heat. The strains shown by a man during work in hot environments are indicated by increased heart rate, rate of sweating, rise in skin and deep body temperatures. Circulatory strain is indicated near the limits of tolerance by a fall of arterial blood pressure. The effects of increasing relative humidity on all these functions become more and more pronounced as dry bulb temperature increases above 28° C as depicted in Fig. 3. (Lind, 1963).

Significant circulatory strain in the working man manifested by heart rate above 140 appeared only when water vapour pressure in the air became so high that heat dissipation by evaporation was limited and rectal and skin temperature rose significantly.

Besides above mentioned physiological strain during work in heat, the total energy cost of a fixed work is also increased. This increase is more with the rise in environmental temperature. This means that thermal stress per se increases energy requirement for work, which may be due to (i) increased demand on blood circulation, (ii) increased activity of sweat glands and (iii) increased body temperature causing higher enzyme activity or biochemical reaction in the body. Moreover, in a hot environment during work, energy is supplied with less supply of aerobic oxygen and with a higher anaerobic component. The increase in anaerobic contribution to the energy supply processes was further confirmed by a significant increase in relative oxygen debt and in the blood lactate level at different work load as shown in Fig. 4.

Thus, a highly significant corelation ($P<0.001$) was observed between oxygen debt contracted with increase in thermal stress. This

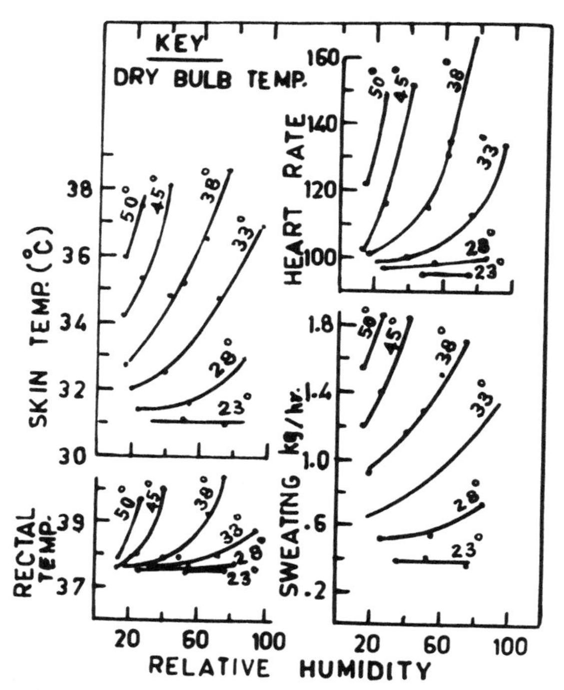

Figure 3: *Physiological responses during work (walking 6.5 km/hr) in hot-dry and humid climate.*

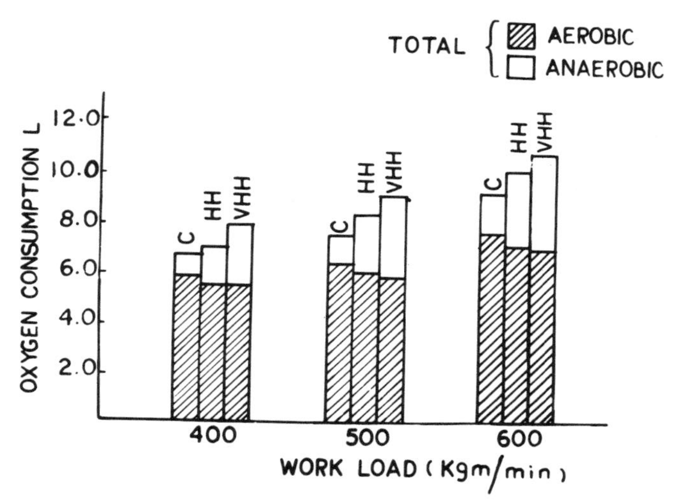

Figure 4: *C - Comfortable temperature*
HH - Hot humid
VHH - Very Hot Humid

shift in aerobic-anaerobic ratio of energy supply is possibly due to diversion of a large amount of blood from the working muscles to the skin for thermoregulation in hot environments. The skin blood flow of 1-2 l/min in hot environment has been reported by Bean and Eichna (1943). If an equivalent proportion of the cardiac output is diverted from the working muscles, they will have to work in reduced blood flow resulting in an increase in the anaerobic metabolism and higher lactate concentration in the blood. A diversion of 2.0 l/min of blood to the skin can lead to approximate reduction in oxygen uptake by 200-300 ml/min during the work.

The significant fall in maximum oxygen uptake capacity has been observed with increase in environmental temperature as shown in Fig. 5.

This reduction in Vo_2 max is, however, higher in hot humid conditions. This indicates that the humid climate has a greater thermoregulatory strain on an individual than in dry heat during heavy work. The fall in Vo_2 max in heat has been reported by Brouha et al. (1961) and Klausen et al. (1967). This reduction in maximum oxygen uptake capacity in heat might have occurred due to reduction in maximum cardiac output. The increased cardiac frequency during work in heat is possibly due to reduced stroke volume as has been reported by Rowell et al. (1965) and Taylor et al. (1955).

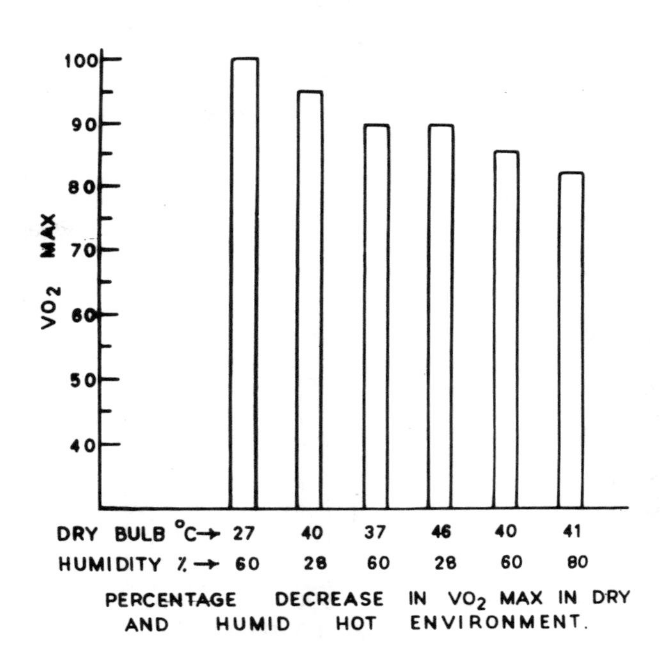

Figure 5: *Percentage decrease in Vo_2 max in dry and humid hot environment.*

PHYSIOLOGICAL RESPONSES IN PROLONGED, CONTINUOUS WORK IN HOT ENVIRONMENT

In our earlier study (Sen Gupta et al., 1984) it has been observed that the duration in continuous physical efforts at different rates under varying climatic conditions decreases significantly as shown in Table 1. It is apparent that the work which could be sustained for more than 90 min in comfortable environment reduced to 23 min only in very hot humid condition. To evaluate the reason for the termination of work, the physiological reactions to continuous work have been analysed. The heart rate responses during the progress of work in different thermal environment has been shown in Fig. 6.

It is observed that the heart rate responses were distinctly much higher in the hot environment than in the comfortable climate. However, the rate of progression of the heart rate towards the target maximum of 180 b/min was much faster in hot humid than in hot dry environments. Thermoregulatory strain of the subjects during different rate of work in different thermal conditions has been evaluated by recording rectal temperature (T_R) and mean skin temperature (T_S), and has been presented in Fig. 7.

TABLE 1: *Duartion (min) in continuous physical activity of varying severity in different hot environments*

Thermal Condition	Rate of Work 400 kg/min	500 kg/min	600 kg/min
1. Comfortable	More than 90 mins	More than 90	Above 90
2. Hot Humid	60 "	60	33
3. Very Hot Humid	45 "	36	23
4. Hot Dry	85 "	75	41

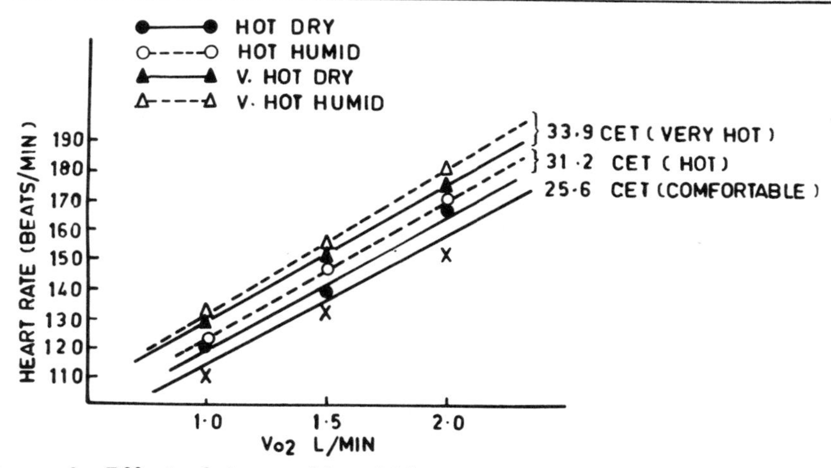

Figure 6: *Effect of dry and humid heat on exercise heart rate.*

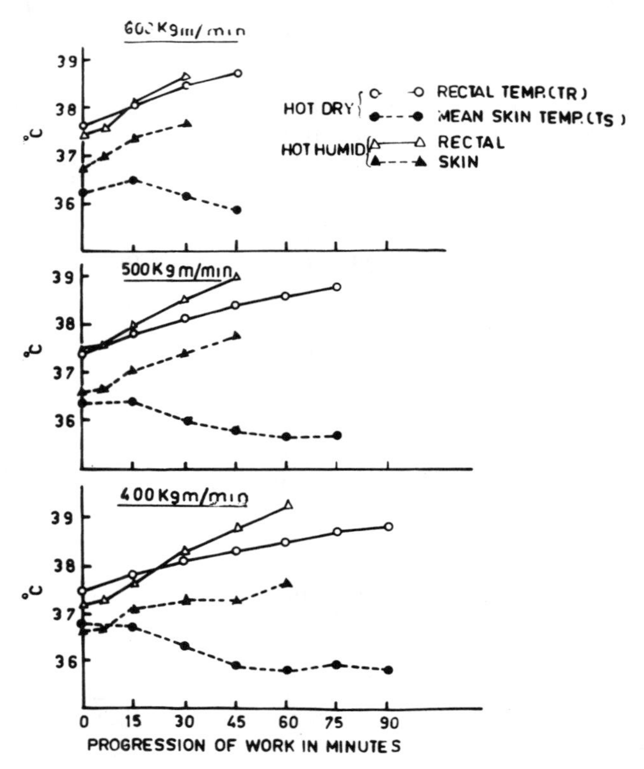

Figure 7: *Comparison of rectal and skin temp. responses during work in hot dry and hot humid conditions.*

It is seen that at comfortable temperature in all the three rates of work T_R maintained a steady state or near steady state condition while the mean skin temperature either maintained or slowly decreased with progression of work. On the other hand, in hot humid condition rectal temperature and skin temperature increased progressively and the gradients between rectal and mean skin temperature narrowed to a great extent. For regulation of the thermal balance of the body, sweating is the principal mechanism of heat dissipation more particularly during work in heat. As such, rate of sweating is a good index of thermoregulatory efficiency. Comparison of the sweating rate during work in hot dry and hot humid conditions is shown in Fig. 8.

It is observed that rate of sweating is comparatively less in dry condition than in humid one. The possible reason for this discrepancy may be in efficiency of cooling in these two conditions. In humid condition, the efficiency is less due to loss of sweat by dripping from the body instead of getting evaporated. It can, therefore, be assumed that the rate of sweating as well as efficiency in evaporation are important criteria in heat regulatory mechanism during work in heat. From the findings of our studies it seems probable that it would be desirable to combine a number of functional aspects into a single index to predict the tolerance of work by man in hot dry and humid conditions, and such an attempt has been made in Fig. 9.

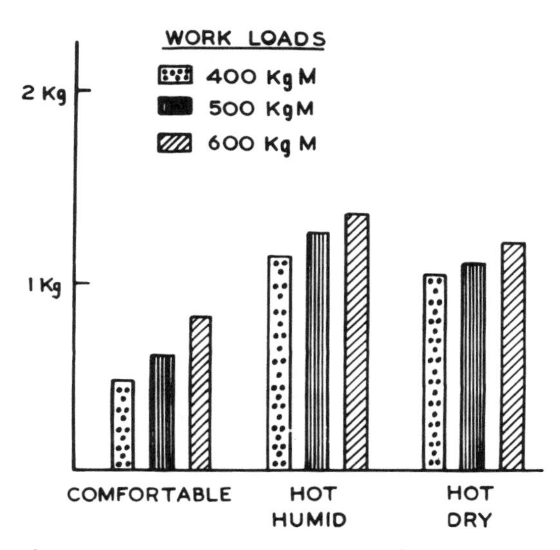

Figure 8: *Rate of sweating during work in different thermal conditions.*

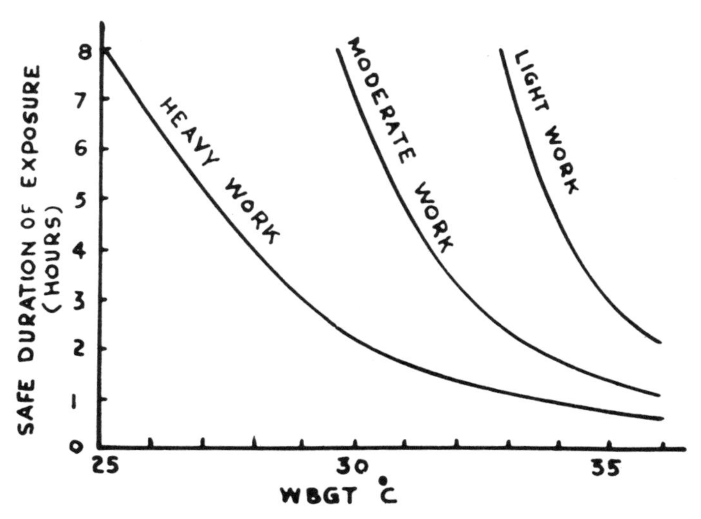

Figure 9: *Safe exposure period of active Indian soldiers in relation to work level and WBGT of environment.*

REFERENCES

ASTRAND, I. (1967): Degree of strain during building work as related to individual aerobic capacity. Ergonomics, 10: 293-303.

ASTRAND, P.O. and RODAHL, K. (1970): Text book of work physiology. Chap. 9. McGraw Hill Company, New York, pp. 303.

BEAN, W.B. and EICHNA, L.W. (1943): Performance in relation to environmental temperature. Fed. Proc., 22: 144-158.

BINK, B. (1962): The physical work capacity in relation to working time and age. Ergonomics, 5: 25-28.

BONJER, F.H. (1962): Actual energy expenditure in relation to the physical working capacity. Ergonomics, 5: 29-31.

BONJER, F.H. (1968): Relationship between physical working capacity and allowable calorie expenditure. In: Int. Colloquim on muscular Exercise and Training. Darmstadt. Germany.

BROUHA, L., SMITH, P.E. jr., DeLANNE, R., and MAXFIELD, M.E. (1961): Physiological reaction of men and women during muscular activity and recovery in various environments. J. Appl. Physiol., 16: 133-140.

COSTILL, D.L. (1972): Physiology of marathon running. J.A.M.A., 22: 1024-1028.

KLAUSEN, K., DILL, D.B., PHILIPS, E.E. jr. and McGRAGOR, D. (1967): Metabolic reactions to work in the desert. J. Appl. Physiol., 22: 292-296.

LEHMAN, G., MULLER, E.A. and SPITZER, H. (1950): Der Kalorienbedarf bei gewerblicher Arbeit. Arbeitsphysiologie, 14: 166-235.

LIND, A.R. (1963): A physiological criteria for setting thermal environmental limits for every day work. J. Appl. Physiol., 18: 51-56.

MICHAEL, E.D., HUTTON, K.E. and HORVATH, S.M. (1961): Cardiorespiratory responses during prolonged exercise. J. Appl. Physiol., 16: 997-1000.

MULLER, E.A. (1950): Ein Leistungs- Puls- Index als Mass der Leistungsfähigkeit. Arbeitsphysiologie, 14: 271-284.

MULLER, E.A. (1962): Occupational work capacity. Ergonomics, 5: 445-452.

PRUETT, E.D.R. (1970): FFA mobilization during and after prolonged severe muscular work in man. J. Appl. Physiol., 29: 809-815.

ROWELL, L.B., RUDENFELDT, H. and FERCH, U. (1965): Studies on the influence of high temperature and high air humidity on the body temperature, respiratory and heart frequency. Wien. Tierärztl.Monatsschr., 52: 436-454.

SEN GUPTA, J., MALHOTRA, M.S. and RAMASWAMY, S.S. (1972): Exercise dyspnoea and work performance at sea level and at high altitude. Ind. J. Physiol. Pharmacol., 16: 47-53.

SEN GUPTA, J., VERMA, S.S., JOSEPH, N.T. and MAJUMDAR, N.C. (1974): A new approach for the assessment of endurance work. Europ. J. Appl. Physiol., 33: 83-94.

SEN GUPTA, J., VERMA, S.S. and MAJUMDAR, N.C. (1979): Endurance capacity for continuous effort in terms of aerobic and anaerobic fraction of oxygen supply. Ind. J. Physiol. Pharmac., 23: 169-178.

SEN GUPTA, J., SWAMY, Y.V. PICHAN, G. and DIMRI, G.P. (1984): Physiological responses during continuous work in hot dry and hot humid environments in Indians. Int. J. Biometeor., 28: 137-146.

TAYLOR, H.L., BUSKIRK, E. and HENSCHEL, A. (1955): Maximal oxygen uptake as an objective measure of cardiorespiratory performance. J. Appl. Physiol., 8: 73-80.

WYNDHAM, C.H., WILLIAMS, C.G. and VON RADHDEN, M. (1982): A physiological basis of the optimum level of energy expenditure. J. Physiol. (Lond.), 195: 1210-1212.

PREDICTION OF ENDURANCE TIME DURING CONTINUOUS WORK IN HEAT

G.P. Dimri and S.S. Verma
(Defence Institute of Physiology and Allied Sciences,
Delhi Cantt-110010, India)

Abstract: - Multiple Regression Equations were fitted to the data of Endurance time, from seven independent physiological and physical variables. Significant variables were identified using the technique of stepwise linear regression analysis.

INTRODUCTION

Endurance may be considered to be the ability of the body to withstand the stresses imposed by prolonged physical activity. Direct determination of human endurance time is quite complicated as it involves the maximum exertion of the subject under investigation. Some workers have attempted indirect estimation of endurance capacity from aerobic capacity (Astrand, 1967; Costill, 1971; 72; 73; Bink, 1962; Bonjer, 1968). Such approaches have limitations in many industrial and military situations where environmental heat is one of the predominant factors and not considered in the above studies. A few workers have studied the relationship of exposure time in heat to the severity of thermal stress by rectangular hyperbolic model (Bell et al., 1969; 1971). However, these authors have taken into account only the thermal stress whereas safe exposure time is affected by other limiting factors like severity of work, cardiorespiratory strains and thermoregulatory imbalance (Wenzel, 1963). The object of this paper is to predict endurance time from number of physical and physiological responses by the procedure of multiple linear regression analysis and identifying the significant factors from among the various physiological variables studied.

MATERIALS AND METHODS

The studies were conducted on 6 young healthy Indian males, residents of hot tropical areas, with a mean age, height and weight of 24.2 yr, 166.2 cm and 52.7 kg respectively. The subjects were given three different rates of workload of 400, 500 and 600 kg.m.min^{-1}

equivalent to 65.40, 81.75 and 98.10 W, under two environmental conditions which were hot humid (D.B.=37° C, W.B.=29.0° C and R.H.=60%) and very hot humid (D.B.=40° C, W.B.=32.5° C and R.H.=60%). The thermal stress was expressed as Oxford Index (WD), according to Lind and Hellon (1957), which was 30.20° C and 33.63° C in hot humid and very hot humid conditions.

The subjects reported to the laboratory in the morning two hours after a light breakfast. They were given one hour rest. Then exercise was given on a mechanically braked bicycle ergometer, at 60 r.p.m. The endurance time for each subject was recorded at each workload and under each environment with the help of stop watch till the attainment of one of the following criteria.

i) The subject can not maintain the assigned rate of work even after best effort;
ii) Attainment of heart rate of 180 beats min^{-1} or above.

Besides, the work was also terminated when the subject could not continue the work rate due to physical fatigue, unsteadiness, exhaustion and feeling of Warmth all over the body. During continuous work the physiological responses of oxygen consumption (Vo_2), pulmonary ventilation (VE), Heart rate (HR), and skin temperature ($\overline{Ts}$) were noted every 15 min and just before the termination of exercise. The subject was weighed on Avery machine sensitive upto 5 g before and after the exercise and sweat rate was determined. For measuring exercise ventilation, the subjects were made to breathe through a low resistance breathing valve (Collins) to a Cowan Parkinson dry gasmeter. The mixed expired air samples were analysed for oxygen and carbondioxide by passing through a calibrated CO_2 analyser (Backman) and O_2 analyser (Servomax controls). The exercise heart rate was recorded in an ECG machine using chest leads. The skin temperatures were monitored from four sites by using a YSI telethermometer and the mean weighted skin temperature ($\overline{Ts}$) was worked out according to the equation of Ramanathan (1964).

REGRESSION ANALYSIS

Multiple linear regression analysis was performed on an IBM 360/44 computer. Multiple linear regression equation with dependent variable Y and seven independent variables (X_1, X_2..........., X_7) is given by (Winer, 1971).

$$Y = a + b_1 x_1 + b_2 x_2 + \ldots\ldots\ldots\ldots + b_7 x_7 \qquad (1)$$

The variables represent:

Y = Endurance time in min.
X_1 = Workload (W) in kg.m.$min.^{-1}$
X_2 = Oxygen consumption (Vo_2) in l min^{-1}
X_3 = Pulmonary ventilation (VE_{BTPS}) in l min^{-1}
X_4 = Heart rate (HR) in beats min^{-1}
X_5 = Mean skin temperature ($\overline{Ts}$) in °C.
X_6 = Sweat rate (SR) in kg. hr^{-1}
X_7 = Thermal stress expressed as oxford index (WD) in °C.

The values of the constant a, b_1, b_2 ...., b_7 were determined by solving simultaneous equations and making use of Inverse matrix of Independent variables. The constants were tested for statistical significance of multiple regression equation. The relevance of partial regression coefficients for significant contribution was tested by analysis of variance technique. The significance of multiple correlation coefficient between observed endurance time and estimated endurance time was tested by F test (Anderson, 1958).

The analysis of multiple regression equation was also performed in stepwise manner (Draper and Smith, 1966) in order to identify the signifcant predictor variables from among the seven independent variables.

One more set of mutiple regression equation was fitted to the data by replacing independent variable workload by relative workload (W/VO_2 max), and once again selection for significant predictor variables was made by stepwise regression technique.

RESULTS

The sample means and standard deviations of the dependent variable (endurance time) and other independent variables are given in Table 1. The correlation coefficient of the measurements with endurance time have been also included in this table and the level of significance for each correlation coefficient has been saparately indicated.

TABLE 1: *Physiological and environmental variables and their correlations with endurance time.*

S.No.	Variables	Unit	Range	Sample Mean	S.D.	Correlation with endurance time
1.	Endurance time	min	15-90	43.97	19.14	
2.	Workload	$kg.m.min^{-1}$	400-600	500	82.81	-0.5949^c
3.	Thermal stress as oxford index	°C	30.20-33.63	31.91	1.74	-0.4872^b
4.	Oxygen consumption	$l\ min^{-1}$	1.167-2.009	1.62	0.22	-0.2306
5.	Pulmonary ventilation	$l\ min^{-1}$	32.92-63.42	47.54	8.84	-0.3088
6.	Heart rate	b p m	156-188	180.50	5.66	-0.3551^a
7.	Skin temperature	°C	35.78-37.98	37.03	0.60	-0.2196
8.	Sweat rate	$kg\ min^{-1}$	0.81-2.40	0.89	0.018	-0.6923^c
9.	Vo_2 max	$l\ min^{-1}$	2.041-2.940	2.56	0.31	-0.4828^b
10.	Relative workload	$kg\ l^{-1}$	136.05-293.97	198.56	42.31	-0.7510^c

Significant levels: a - $p < 0.05$
b - $p < 0.01$
c - $p < 0.001$

Two multiple regression equations with seven independent variables, one with workload and a second with relative workload were obtained. The multiple regression equation with workload gave a multiple correlation coefficient (R) 0.84 between observed and estimated endurance time. When workload was replaced with relative workload, the multiple correlation coefficient (R) increased to 0.88. Since all the variables were not found significant as per the relevance of partial regression coefficients occurring in the two multiple regression equations, the method of stepwise regression was followed to select variables which contribute significantly. The list of prediction equations for determining endurance time from variables inclusive of workload have been given in Table 2 together with multiple correlation coefficient (R), the standard error of estimate for each equation and the standard error of partial regression coefficients. Out of seven variables selected for the study only four variables were found significant predictors of endurance time. These variables were (i) workload, (ii) thermal stress quantified as Oxford index, (iii) mean skin temperature and (iv) sweat rate. The multiple regression equation constructed with these four variables accounted for 70% of variation in endurance time. It has been observed that multiple correlation coefficient was highly significant ($p<0.001$) for each of the equations in Table 2.

The partial regression coefficients were tested for the significance of their contributions by analysis of variance technique and they were

TABLE 2: *Multiple linear Regression Equations for predicting endurance time from physiological and environmental variables in stepwise manner. (The set of variables includes workload).*

S.No.	Regression Equation Endurance time (Y)	r or R	Standard error of estimate	Standard error of partial regression coefficients			
				b1	b2	b3	b4
1.	$91.90620-33.45541\ X_1^c$	0.69225	14.01362	5.98124			
2.	$120.92969-25.62611\ X_1^c -0.08048\ X_2 b$	0.75777	12.86160	6.20157	0.02966		
3.	$229.84035-14.96063\ X_1^a -0.10421\ X_2^c-3.51956\ X_3^a$	0.79898	12.03709	7.32999	0.02949	1.47734	
4.	$-122.29785-12.76019\ X_1^a -0.11196\ X_2^c-7.16922\ X_3^c+ 12.67361\ X_4^a$	0.83730	11.11903	6.82570	0.02741	4.97007	

X_1 - Sweat rate
X_2 - Workload
X_3 - Thermal stress expressed as Oxford index (WD)
X_4 - Skin temperature

r - Coefficient of correlation
R - Multiple correlation

Significance Levels:
a - $p < 0.05$
b - $p < 0.01$
c - $p < 0.001$

found to give significant contributions individually and collectively for the prediction of endurance time.

The list of prediction equations for determining endurance time from variables inclusive of relative workload have been given in Table 3 together with multiple correlation coefficient (R), the standard error of estimate for each equation and the standard error of partial regression coefficients. It is observed that the new multiple regression equation constructed with three significant predictor variables inclusive of relative workload accounts for 78% variation as compared to earlier equation accounting for 70% variation in the estimation of endurance time. Thus it shows the superiority of relative workload over workload as a significant predictor of endurance time.

DISCUSSION

The present study was planned to estimate endurance time in terms of the physiological variables considered as safe with limitation of physiological strain. The criteria of overstrain from exposure to heat for defining the limits of endurance while working in heat has been suggested by Wyndham et al. (1965), Lind et al. (1963) and the expert committee seeking international accord on proper limits for occupational exposure to heat (WHO Tech Rep Series no. 412, Geneva 1969). The physiological criteria of heart rate touching 180 beats min^{-1} was followed in hot humid and very hot humid environments in the present study for the termination of exercise.

TABLE 3: *Multiple linear Regression Equations for predicting endurance time from physiological and environmental variables in stepwise manner. (The set of variables includes Relative workload).*

S.No.	Regression Equation Endurance time(Y)=	r or R	Standard error of estimate	Standard errors of partial regression coefficients b1	b2	b3
1.	$111.41968-0.33968\,X_1^{c}$	0.75094	12.82345	0.05123		
2.	$126.10045-0.25332\,X_1^{c}-22.21405\,X_2^{c}$	0.85950	10.07471	0.04424	4.72706	
3.	$189.62074-0.25880\,X_1^{b}-16.74280\,X_2^{b}-2.20184\,X_3^{a}$	0.87576	9.66268	0.04252	5.31799	1.11862

X_1 = Relative workload
X_2 = Sweat rate
X_3 = Thermal stress expressed as Oxford Index (WD)

a - $p < .05$
b - $p < .01$
c - $p < .001$
r - Coefficient of correlation
R - Multiple correlation.

The various indices developed by other workers for assessing endurance capacity are of limited utility because some of them have studied the problem with respect to the severity of workload alone (Bink, 1962; Bonjer, 1968; Astrand, 1967; Muller, 1962) where as others with respect to heat load only (Bell and Walters, 1969; Bell and Crowder, 1971). In the present study we have studied endurance time not only with workload and thermal stress but also with severity of various physiological responses as emphasized by Wenzel (1963). Out of seven independent variables studied, four variables only were found to make significant contributions. The findings are in agreement with Consolazio et al. (1978) and Libert et al. (1978) who found Sweat rate and skin temperature as direct index of physiological strain.

Further Vo_2 max is reported to be the best measure of physical work capacity (Astrand, 1960; Dimri et al., 1981). We have observed a significant association between endurance time and Vo_2 max in this study ($r = 0.48$, $p < 0.01$). The severity of a workload alters due to the oxygen carrying capacity of blood while working in heat (Consolazio et al., 1963; Sen Gupta et al., 1977, Dimri et al., 1980). It was therefore considered better to replace workload by relative workload for the prediction of endurance time. The same was tried in multiple regression equation and three variables namely relative workload, sweat rate and thermal stress were found to make significant contributions.

Stepwise linear regression analysis has been utilised in this investigation of 36 observations which may not be sufficient for the prediction equations to be considered as generalised approach. A future study with larger sample size may prove more informative. The observations may be considered as independent owing to different physiological conditions of the subjects on different days at different work loads (Bell and Crowder, 1971). In the study of Bell and Walters (1969), the safe exposure time in a step up test on stool has been reported to decrease from 52.9 min to 5.9 min with increasing severity of heat. A similar trend of decrease in the endurance time has been found in the present investigation. The multiple regression equation constructed with four significant predictors of workload, thermal stress, mean skin temperature and sweat rate account for 70% variation in endurance time. On replacing workload with relative workload, the multiple linear regression equation gives better prediction by accounting for 78% variation in the endurance time. This approach is likely to be useful in such industrial and military situations where similar workload and environmental conditions prevail.

ACKNOWLEDGEMENTS

The Authors' sincere thanks are due to Gp Capt K.C. Sinha, Director, D.I.P.A.S. and Dr. J. Sen Gupta, Deputy Director D.I.P.A.S. for their constant encouragement of the work and permission to publish the paper. Thanks are also due to Shri Omparkash for his secretarial assistance.

REFERENCES

ANDERSON, T.W. (1958): An introduction to multivariate statistical analysis. John Wiley and Sons, New York.

ASTRAND, I. (1967): Degrees of strain during building work as related to individual aerobic capacity. Ergonomics 10: 293-303.

BELL, C.R. and WALTERS, J.D. (1969): Reactions of men working in hot and humid conditions. J. Appl. Physiol. 27: 684-686.

BELL, C.R. and CROWDER, M.J. (1971): Durations of safe exposure for men at work in high temperature environments. Ergonomics, 14: 733-757.

BINK, B. (1962): The physical working capacity in relation to working time and age. Ergonomics 5: 25-28.

BONJER, F.H. (1968): Relationship between physical working capacity and allowable calorie expenditure. In: International colloquium of muscular exercise and training. Darmstadt, Germany.

CONSOLAZIO, C.F., MATOUSH, L.R.O., NELSON, R.A., TORRES, J.B., and ISSAC, G.J. (1963): Environmental temperature and energy expenditure. J. Appl. Physiol. 18: 65-68.

COSTILL, D.L., BRANAM, G., EDDY, D. (1971): Determinate of marathon running success. Int. Z. Angew. Physiol. 29, 249-254.

COSTILL, D.L. (1972): Physiology of marathon running. J.A.M.A. 22, 1024-1028.

COSTILL, D.L., THOMSON, H., and ROBERTS, E. (1973): Fractional utilization of the aerobic capacity during distance running. Med. Sci. Sports, 5, 248-252.

DIMRI, G.P. MALHOTRA, M.S., SEN GUPTA, J., SAMPAT KUMAR, T. and ARORA, B.S. (1980): Alterations in aerobic-anaerobic proportions of metabolism during work in heat. Eur. J. Appl. Physiol. 45: 43-50.

DIMRI, G.P., SAMPAT KUMAR, T., PILLAY, P.B.S., ARORA, B.S., SINGH, H., SRIDHARAN, K., SEN GUPTA, J. and NAYAR, H.S. (1981): Physical work capacity of Indian adolescent boys in Sainik and Military schools. Indian J. Med. Res. 74: 610-616.

DRAPER, N.R., and SMITH, R. (1966): Applied Regression Analysis - John Wiley and Sons, New York.

GONALEZ, R.R. BERGLUND, L.G., and GAGGE, A.P. (1978): Indices of thermoregulatory strain for moderate exercise in the heat. J. Appl. Physiol. 44, 889-899.

LIBERT, J.P., CAUDAS, V., and VOGT, J.J. (1978): Effect of rate of change in skin temperature on local sweating rate. J. Appl. Physiol. 47, 306-311.

LIND, A.R. (1963): A physiological criteria for setting thermal environmental limits for every day work. J. Appl. Physiol. 18: 51-56.

MULLER, E.A. (1962): Occupational work capacity. Ergonomics, 5: 445-452.

RAMANATHAN, N.L. (1964): A new weighing system for mean surface temperature of human body. J. Appl. Physiol. 19: 531-533.

SEN GUPTA, J., DIMRI, G.P. and MALHOTRA, M.S. (1977): Metabolic responses of Indians during submaximal and maximal work in dry and humid heat. Ergonomics, 20: 33-40.

WENZEL, H.G. (1963): Studies on the question of defining tolerable and intolerable heat load for man during physical work lasting several hours. Arch. Ges. Physiol. Menschen u. Tiere, 278-96.

WHO TECH Report series No. 412, (1969): Health factors involved in working under conditions of heat stress. Geneva.

WYNDHAM, C.H., STRYDOM, N.R., MORRISON, J.F., WILLIAMS, C.J., BREDELL, G.A.G., MARITZ, J.S. and MUNXRO, A., (1965): Criteria for physiological limits for work in heat. J. Appl. Physiol. 20: 37-45.

A METHOD FOR EVALUATING PHYSIOLOGICAL HEAT STRESS IN INDIA

A. Chowdhury, S.S. Singh
(Meteorological Office, Pune)
and
H.R. Ganesan
(Mukund Nagar, Pune)

Abstract: - In this study, an index has been devised incorporating non-meteorological parameters like clothing, metabolic heat rate, etc. and meteorological parameters i.e. temperature and humidity, to assess thermally comfort conditions. For this purpose daily temperature and the vapour pressure for 60 selected stations, have been utilised. Mean monthly frequencies when the index assumes values greater than 0.75 and less than 0 in different class intervals have been analysed and discussed.

INTRODUCTION

Physiological strain on human beings is normally caused by climatic conditions. A number of empirical relationships are available in the biometeorological literature on thermal stress on a normal human with sedentary activity. One important result emerging from this work is that in tropical regions both temperature and humidity contribute significantly to the physiological stress.

In this paper an index developed earlier by Lee and Hanschel (1963) has been used to delineate areas of climatic stress for different season over India.

DATA UTILISED

Heat being the main causative factor of the stress, day time temperature has been used in the study. Over large parts of India the maximum temperature is normally attained between 2 to 3 p.m. (i.e. 1400 to 1500 hrs). Because of the easy availability of temperature and humidity at 1430 synoptic hour, in this study, data for this particular hour is used. The study utilises daily data from 1969-73 for 60 stations.

DETERMINATION OF THE INDEX

The physiological stress index called Relative Stress Index (RSI) as given by Lee and Hanschel (l.c.) is

$$RSI = \frac{M (I_{cw} + Ia) + 5155 (ta - 35) + RIa}{7.5 (44 - Pa)}$$

Where,

M = Metabolic rates in k cal m^{-2} hr^{-1}
ta = Air temperature in °C.
R = Mean radient energy incidence in k cal m^{-2} hr^{-1}
Ia = Insulation of air in clo units.
I_{cw} = Insulation of moist clothing in clo units.
Pa = Vapour pressure of air in mm Hg.
Substituting, M = 100K cal m^{-2} hr^{-1}
(equivalent to walking at 2mi-hr^{-1})
I_{cw} = 0.4 clo (standard business suit)
Ia = 0.4 (ambient air movement 100 ft min^{-1} air movement caused by body movement 100 ft min^{-1})
R = 0 (wall and air temperature considered equal)
the equation reduces to substituting these values,

$$RSI = 10.7 + 0.74 \frac{(ta-35)}{(44-Pa)}$$

In the present study the mean monthly values of the index have been obtained, analysed and discussed. Frequency of the index assuming values less than 0 and greater than 0.75 have also been obtained and discussed. Four representative months viz. January, April, July and October have been selected in the study. Besides, index values for May, the hottest month over most parts of the country, have also been analysed.

RESULTS AND DISCUSSION

The mean monthly index:

January: The index is negative or zero in Punjab, Himachal Pradesh, Jammu and Kashmir, Haryana and West Uttar Pradesh. In Jammu and Kashmir in particular, the stress is felt much. This is mostly due to low temperature and this index assumes values lower than 0.20 over this state. In the northeast region in Mizoram the index values reach zero. The largest positive values are seen in the south extreme peninsula and coastal Andhra Pradesh where the stress index exceeds 0.20.

April: In Jammu and Kashmir the stress index continues to be negative though small. Low values persist over northeast region and around Bangalore. The values in the east coast are markedly large particularly around Visakhapatnam and Nellore. These areas experience the maximum amount of stress.

May: The effect of the winter largely disappears from Jammu and Kashmir though the strain values are still low (less than 0.2). The summer conditions, however, are reflected in the rest of the country with substantial rise in the index value, compared to April. The stress on the east coast further aggravates (Fig. 1). Heat stress is also felt over the Konkan and coastal areas of Gujarat. The hilly areas continue to have rather comfortable climatic conditions.

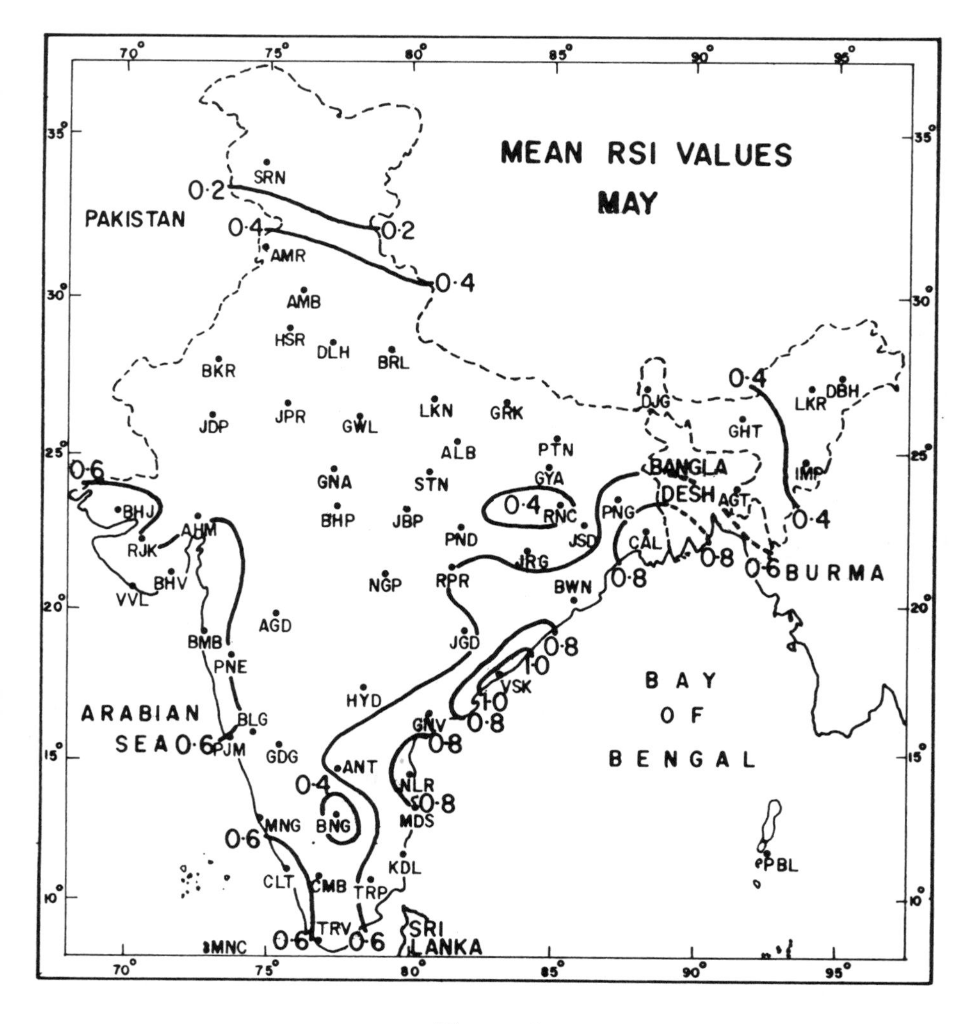

Figure 1.

July: With the onset of monsoon conditions the strain value decreases in the peninsular region (Fig. 2). The belt of extreme stress now shifts to northwest India. Stress on the coast particularly around Visakhapatnam continues unabated though with diminished intensity.

October: During October the falling trend in RSI commences over Jammu and Kashmir. The index values however are still less than 0.2. Low values also persist over some pockets around Bangalore and in Bihar plateau. Over the coastal areas the values are large i.e. more

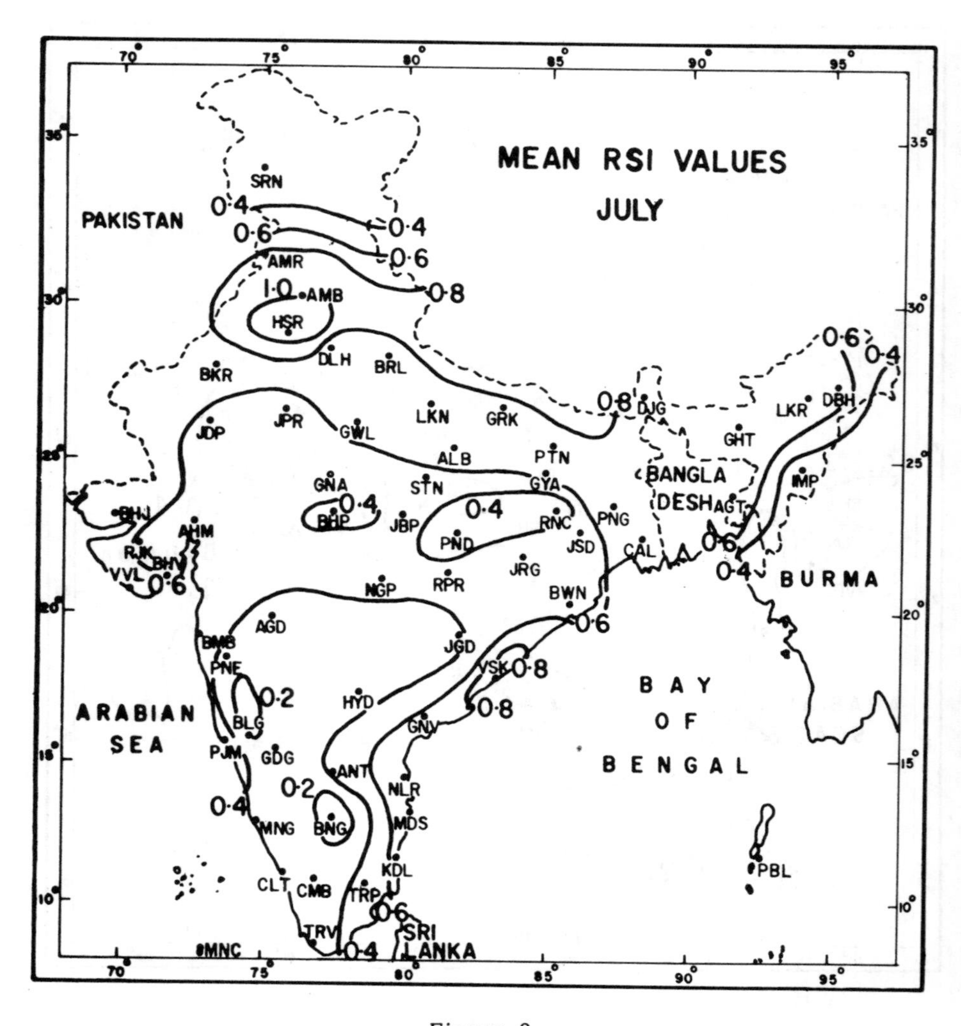

Figure 2.

than 0.4. The largest value i.e. 0,6, is located over the Saurashtra coast.

The broad conclusions that emerge from this analysis are:

1) For most part of the year the thermal stress is not much felt over Jammu and Kashmir as also hilly terrain.
2) Coastal areas, particularly the east coast are prone to high load of thermal stress throughout the year. In particular around Visakhapatnam the stress appears oppressive round the year.

FREQUENCY DISTRIBUTION

(a) RSI $\leqslant$ 0

January: Above approximately 23°N the RSI is invariably less than or equal to zero during this month while south of 23°N, normally there is hardly any occasion when the index is less than zero. Over the northwest India, in more than 20% of the occasions (in Jammu and Kashmir it is on more than 30%) the values are less than zero. Negative values are mostly contributed by the low temperatures in north India.

April/May: Except Jammu and Kashmir where in a large number (about 20 days in April and 10 days in May) of the occasions, the values are still less than zero, over rest of the country the values are generally above zero, during April and May.

July: During July nearly over whole of the country there is hardly any day when RSI $\leqslant$ 0.

October: In October because of the onset of winter conditions, in only Jammu and Kashmir on about 10 days the RSI is zero or less. Over rest of the country the RSI is generally above zero on most days of the month.

(b) RSI > 0.75

January: There is no occasion in any part of the country where the RSI assumes values more than 0.75 during January.

April: Over Bihar, northeast region, north Madhya Pradesh, Uttar Pradesh and NW India and also hilly areas of the peninsula during April the RSI generally does not rise more than 0.75. Over the peninsula on a few occasions, the index exceeds 0.75. Along the coastal areas of Bengal, Andhra Pradesh and north Tamil Nadu, occasions having RSI greater than 0.75 exceed 10%.

May: During May though over the country as a whole RSI exceeds 0.75, on most of the days, in parts of east Rajasthan and adjoining Madhya Pradesh, on lee side of the Ghats in Madhya Maharashtra and also Jammu and Kashmir and extreme NE India, on no occasion the value exceeds 0.75. These areas are thus comfortable even during the height of summer. Prolonged strain as in the month of April is also felt in the east coast.

July: With the onset of the monsoon conditions, in most part of the peninsula (except the east coast) the frequency is not large. However in north India, the strain is of fairly larger duration and in 15-20 days in the month, the RSI exceeds 0.75.

October: In October the pattern is similar to that observed in January. But in the Konkan coast and coastal Tamil Nadu on a few occasions the values exceed 0.75.

SUMMARY

An attempt has been made in this work to identify and demarcate areas having high load of physiological stress due to weather factors. The degree of discomfort and its frequency distribution have also been obtained and discussed.

ACKNOWLEDGEMENT

The authors are thankful to Dr. A.K. Mukherjee for encouragement. Thanks are also due to Shri S.K. Dey, Smt. Mhasawade and Kum. P.G. Gore for collection and computation of data and to Smt. Chandrachood for typing.

REFERENCES

LEE, D.H.K. and HANSCHEL, A. (1963): Evaluation of thermal environment in shelters. TRS US Dept. of Health, Education and Welfare Cincinnati, Ohio, p. 58.

A MODEL ON OUTGOING WATER BUDGET FOR THE HUMAN BODY IN TROPICAL CONDITIONS

A.K. Mukherjee, A. Chowdhury and S.S. Singh
(Meteorological Office, Pune, India)

Abstract: - The creation and maintenance of water need is the most essential priority for the human body. Therefore, the climatology of the outgoing water budget from the human body becomes more important. In the present study, confined to the Indian region, an attempt has been made to provide the climatology of the outgoing water budget from the human body in terms of respiratory *and non-respiratory evaporative losses.*

INTRODUCTION

The water requirement of a human body plays an important role in the energy budget of man. Systematic studies of energy budget in terms of heat viz. the amount of heat absorbed and lost by the human body have been made, mostly in other countries (Fanger, 1972; Gagge et al., 1971; and Burt et al., 1982). Such studies do not appear to have been undertaken under tropical conditions. In the present study, an attempt has been made to compute the water budget of the human body in Indian tropical conditions.

Water loss from the human body consists of loss of water vapour to the atmosphere due to breathing, called respiratory loss and the non-respiratory evaporative loss. The respiratory loss depends on the difference in mixing ratios of the expired and ambient air and the rate of breathing. This loss is controlled by two factors, namely, the availability of water in the body and the transport of the vapour. The non-respiratory heat loss is a function of the fraction of the body wetted by sweat glands and the evaporative power of the atmosphere. In this study these losses have been evaluated.

METHODS

Monthly normals of the maximum temperature and vapour pressure for 60 fairly well distributed stations have been used. The temperature data were collected from climatological Tables of observations in India, published by the India Meteorological Department (1931-60). From the wet bulb and dry bulb temperatures, vapour pressure has been calculated using hygrometric tables. In calculating the evaporative power of the environment in the absence of appropriate data, the potential evaporation computed by Rao and George (1971) has been used to compute the non-respiratory losses.

COMPUTATION OF THE LOSSES

(a) Respiratory loss:

According to Mc Cutchan and Tayler (1951), if the air in the lungs is saturated and is at 37° C, the evaporative loss E_v is given by

E_v = $1.725 \times 10^{-3} \times M (33 - P_c)$ Watt
Where M = Metabolic heat, given by
M = $GB_{S'}$
G = metabolic rate (assumed 116 w/m^2)
$B_{S'}$ = Body surface area.
= $0.202 (wt)^{.425} (Ht)^{.725}$ m^2
Wt and Ht being weight of the body (kg) and height (m) respectively are assumed as 55 kg and 1.55m.
P_c = Partial pressure of water vapour (mb) substituting these values the E_v reduces to
E_v = $11.6534 \times (33-P_c)$ g/day.

(b) Non-Respiratory evaporative losses:

Gagge, Stolwijk, and Nishi (1971) showed that the water vapour flux through the skin E_s is dependent only upon the evaporative capacity of the environment.
E_s = $0.06\ E_{max}$ Watt
or E_s = $2.9497 \times E_{max}$ g/day.

Where E_{max} = evaporative power of the environment (W).
(c) The sensible heat sweating E_{sw} is given by $E_{sw} = .94\ W'\ E_{max}$
Where W' = proportion of body wetted by sweating.
If Ts = mean daily maximum temperature, W' assumes the following values:
W' = $(Ts-33)^2 / 16$ $33° \leq Ts \leq 37°$
= 0 $Ts < 33°$
= 1 $Ts > 37°$
Thus, $E_{sw} = 46.2121 \times E_{max} \times W'$ g/day

In the computations, only the influence of skin temperature (which has been taken as equal to the daily mean maximum temperature) is considered in determining W'.

In the following paragraphs geographical distribution of these biometeorological parameters are discussed for typical months of each season.

RESULTS AND DISCUSSION

January:

E_V: Respiratory loss is found to increase with latitude. At Srinagar it is 125 g while over the coastal areas in the south it is as low as about 50 g. Higher values are seen over the hilly areas of extreme north and western Rajasthan. In the peninsula except the coast, the respiratory loss, by and large, is less than 100 g. A ridge in the isopleths is seen running from Anantpur to Madurai. The modulating influence of the sea is evident from the low E_V values on the peninsula coast. The areas of high losses are those with high topographic altitude. This is because at these locations, the dew point is less and the partial vapour pressure (P_c), is also low.

E_S: The insensible sweating is a direct function of evapotranspiration. It decreases in a reverse way to that of the respiratory loss. The minimum value of about 20 g is in the extreme north while the maximum (about 140 g) is located in the Tamilnadu and Kerala coast. Above 20° N the isopleths are approximately parallel to latitudes.

E_{SW}: The insensible loss is dependent on the daily maximum temperature and the evapotranspiration. During January, because of low maximum temperature and low level of the humidity the E_{SW} is neglible.

T_E: Total water loss is the sum of the respiratory and non-respiratory losses from the human body. During January, there is not much variation in the T_E values and it ranges from 125 g to 225 g as in case of respiratory loss. Minimum value is in the foot hills of Himalaya. Over most part of the country, the values are fairly large (viz., 150 to 200 g). The largest values i.e. exceeding 200 g is seen over central parts of the peninsula as also over the Thar desert.

April:

E_V: Over the central part of the country the values continue to exceed 100 g in April. Elsewhere the respiratory losses are low. There is a slight decrease in the E_V values compared to the winter months. This is because advance of summer monsoon conditions results in increase of dew point and consequently of partial vapour pressure. The pattern of isopleths, generally is same as seen in winter, at south of 15° N with the ridge seen in January pattern which not only persists but now extends upto Aurangabad. Low values are still located over the coastal areas of south peninsula. A few pockets of higher values are around Jaipur (120 g) and Bhopal (120 g). A strong gradient in the isopleths also develops around the east coast from coastal Andhra Pradesh to coastal Bengal suggesting that the inland penetration of the sea effect is not much.

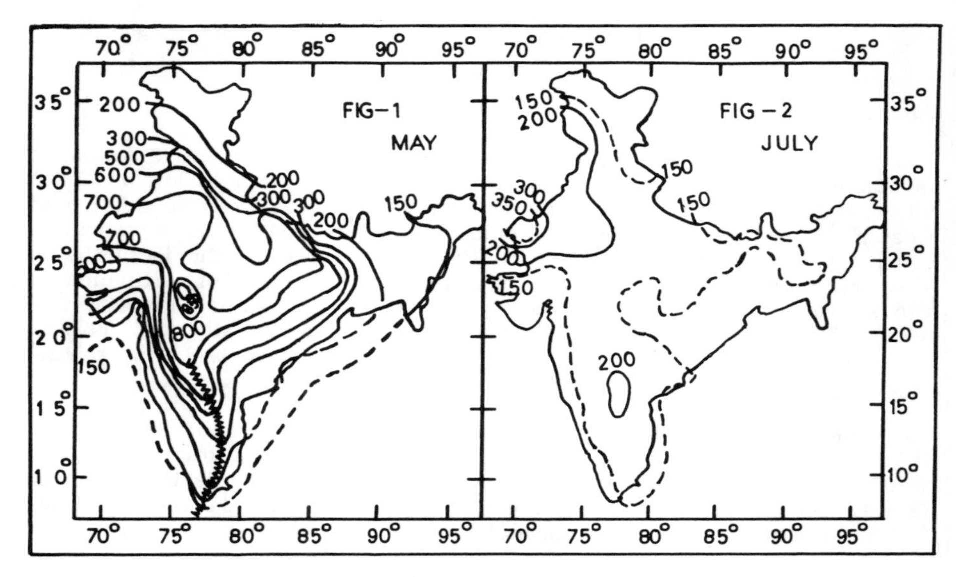

Figure 1 and 2
Daily distribution of total outgoing moisture loss for human body (in g)

E_S: With the advance of summer conditions, E_S increases throughout the country except the Kerala coast. The high altitude regions (viz. Jammu and Kashmir, sub-Himalayan West Bengal and the north-east India) continue to experience low insensible sweating compared to other parts. Strong gradient in these regions also prevails.

E_{SW}: The area under sensible sweating is considerably large because of increased summer conditions, with the core having values more than 300 g located over Vidharbha, Marathwada and adjoining areas of Telengana and Karnataka. Values of comparable magnitude are also observed over central Rajasthan and Bihar Plateau.

T_E: With the advance of summer conditions T_E increases considerably except the Kerala coast where a fall of 10 to 20 g is observed, due to the commencement of pre-monsoon thunderstorm activities. T_E value exceeding 600 g is seen over interior Maharashtra. Another area possessing such high values is seen around Jodhpur and Gaya. Over the coast as also over the hilly terrain, values are considerably less.

May:

E_V: This is the hottest month of the year over most parts of the country. The falling trend in the respiratory loss continues. The highest value of about 100 g is confined to Jammu and Kashmir and over central parts. The lowest value of 10 g is observed over the coastal Andhra Pradesh, with a secondary belt (20 g) located over Kerala coast.

E_S: Because of intense heat, we expect the insensible sweating to be maximum during May. The values throughout the country generally increase. The gradient over the country in the isopleths also increases. In all other respects the pattern resembles that observed in April.

E_{SW}: The pattern of E_{SW} observed in April more or less persists in May. However, the core in the peninsula observed in April shifts northwards and is now seen over west Vidarbha and adjoining Madhya Pradesh. Over large parts of the country the sweating values are about 300 g or more. But over coasts and high elevated regions, the sensible sweating is negligible, as in April.

T_E: T_E further decreases over the west coast. However, over the east-coast the T_E value generally rises. The belt of maximum T_E shifts northwards and is now located over southwest Madhya Pradesh. A strong gradient in the isopleths prevails over the peninsula. Low values continue to be located over the hills and Kerala coast.

July:

E_V: By this month, the monsoon is fully established over the country. As a result E_V generally decreases. The pattern consequently becomes diffuse. High elevated areas in the north and the east have large values (exceeding 60 g).

E_S: As in case of evaporative loss the insensible sweating also decreases in the north. The decrease is most conspicuous in Rajasthan where the decrease is between 100 to 150 g compared to May. But values over western Rajasthan continue to be large.

E_{SW}: In July, due to the overcast skies there is hardly any sensible sweating in any part of the country except north-west India, where some sweating loss occurs.

T_E: The pattern of T_E undergoes a drastic change in July. T_E values all over the country, specially, over north India, generally falls. In Rajasthan itself, the decrease is between 300 to 500 g from what was observed in summer. The decrease in T_E values is very little in north-east India and coastal areas. Over most of the parts of the country T_E is between 150 to 200 g.

October:

E_V: Consequent to the fair weather conditions setting in over the country, E_V values in October rises compared to those seen in July.

Coastal areas of peninsula have low values. Higher values persist over the foot hills of Himalaya and other elevated areas.

E_S: There is generally a decrease in E_S values all over the country except the Kerala coast, where there is a slight increase in July. The minimum values are seen over foot hills of Himalayas as also around Pachmarhi in the Satpura range. High values persist over Rajasthan, south-east Tamilnadu coast and some central parts of peninsula.

E_{SW}: Pattern of sensible sweating during October more or less corresponds to that in July. Only in the western parts of country it is like that of July. Rest of the country have zero E_{sw} values.

T_E: The pattern of T_E becomes diffused in October though the falling trend continues. West Rajasthan continues to be the area with maximum loss (approximately 250 g). Losses elsewhere are less than 200 g. On the west coast they are even less than 150 g.

The broad conclusions that emerge from this study are:

(i) Throughout the year large respiratory water loss from the human body generally occurs over the foot hills of Himalayas, high elevated regions and central parts of north India whereas low values are observed over the coastal areas.

(ii) Respiratory loss has a latitudinally increasing trend while non-respiratory losses show a reverse trend.

(iii) In most parts of the country, except the foot hills of Himalaya and the coastal areas, sensible sweating is felt during the summer season.

(iv) Through the year, maximum total outgoing water loss is observed in central parts of north India and the peninsula.

(v) With the advance of summer conditions the gradient in the isopleths of all type of losses increases, particularly over the east coast.

ACKNOWLEDGEMENT

The authors express their grateful thanks to Shri H.M. Chaudhary, Additional Director General of Meteorology (Research) for providing facilities to complete this study. They are also thankful to S/Shri Godbole for preparing the computer programme and S/Shri D.D. Joshi, O.P. Keswani, P.S. Raut, Smt. M.M. Dandekar and Smt. M.S. Chandrachood for data collection and typing the manuscript.

REFERENCES

BURT, J.E., O'ROURKE, P.A., and TERJUNG, W.H. (1982): Relative influence of urban climates on outer human energy budgets and skin temperature. Int. J. Biomet., 26: 113-123.

DUBOIS, D. and DUBOIS, E. (1916): A formula to estimate the approximate surface area if height and weight be known. Arch. Intern. Med., 17: 863-871.

FANGER, P.O. (1972): Thermal comfort: Analysis and application in environmental engineering. Mc Graw-Hill, New York.

GAGGE, A.P., STOLWIJK, J.A.J., and NISHI, Y. (1971): An effective temperature scale based on a simple model of human physiological regulatory response. Trans. Amer. Soc. Heat., Refrig. Air Cond. Eng., 77: 247-257.

MANI, A. (1980): Handbook of solar radiation. Allied publishers private Ltd., New Delhi, pp. 1-4.

RAO, K.N., GEORGE, C.J., and RAMASASTRI, K.S. (1971): Potential evapotranspiration (PF) over India. Pre-pub. Sc. Rpt. No. 136 Ind. Met. Deptt.

EFFECTS OF EXPOSURE TO HIGH AMBIENT TEMPERATURE (44° C) ON AUTONOMIC BALANCE IN HUMAN ADULTS UNDER DIFFERENT CLIMATIC CONDITIONS

L. Rai*, P.L. Ahujarai**, and S.O.D. Bhatnagar*
(School of Environmental Sciences,
Jawaharlal Nehru University, New Delhi-110067)

Abstract: - Effect of exposure to heat stress (44° C) for 75 min on reactivity of individuals in terms of quantitative shift in autonomic balance ($\bar{A}$ score) and autonomic response pattern (ARP) were studied on thirty male subjects. The results revealed that parasympathetic (PNS) predominant individuals did not show significant change in $\bar{A}$ scores at 44° C whereas sympathetic (SNS) predominant and those manifesting balanced autonomic state tended to shift towards the SNS direction in ambient heat. The pattern analysis indicated that at 44° C the individuals were characterised by sparcity of salivation, fast heart rate, high blood pressure and high palmar and volar skin conductances, and large pupillary diameter. The overall shift in the $\bar{A}$ score as measured by using the 'Normative Regression Equation' could be regarded as an index of physiological responsiveness to heat 'stress'. The magnitude of shift in the $\bar{A}$ score could be used as an index of assessing overall reactivity to a specific stress.

INTRODUCTION

Autonomic nervous system (ANS) is involved in the regulation and execution of visceral responses such as vasoconstriction (Cannon, 1937; Barcroft and Edholm, 1946), vasodilation (Lewis and Pickering, 1931; Sarnoff and Simeone, 1947), Piloerection, sweating (List and Peet, 1938; Adolph and Molnar, 1946; Thomas and Korr, 1952), shivering (Uvnas, 1954; Birziz and Hemingway, 1957) and salivation. Not only is there an involvement of ANS in these responses, but the responses requiring the redistribution of blood in the body which gets markedly evoked by thermal factors may also include participation of ANS as manifested by inhibition of some and augmentation of the other visceral responses. Exposure to heat stress demands intensive activity of the autoregulative mechanisms through the participation of ANS and endocrines. The latter have been demonstrated to be dependent on changing ambient temperature which may or may not be directly

---

Present Addresses:
* Central Research Institute for Yoga, New Delhi-110001, India.
** Department of Environment, Bikaner House, New Delhi-110011, India.

interacting with the individual's microenvironment (Ahuja and Sharma, 1971; Bajaj et al., 1973), and the endocrinal changes in turn influence the susceptibility of ANS. The present work was therefore, initiated with the aim of determining ANS susceptibility in terms of autonomic scores ($\bar{A}$) and autonomic response patterns (ARP) in an individual and then finding reactivity changes in him following the exposure to high ambient temperature (T_a) of 44° C. For doing so, 30 adult human male subjects were studied during Winter (outdoor climate: 21.34° C, 72.35RH) and during Summer (outdoor climate: 40.83° C and 29.34RH) in a room where temperature was maintained at 27° C, and then exposed on the same day in a hot room maintained at 44° C for 75 min.

METHODS

Experimental procedures for measuring functional ANS status using a seven test battery: - The regression equation for assaying ANS status was derived from a factorial study of 110 normal healthy adult males in the age range of 20 to 30 years (Rai, 1978). The means and standard deviations (SD) of that sample were employed for obtaining standard estimates of this equation called scores of autonomic balances or $\bar{A}$ scores for the subjects of the present experiment. The equation is:

$$(\bar{A}) = 0.35\ T_{(SO)} + 0.03\ T^{a}_{(PD)} + 0.28\ T^{a}_{(DBP)} + 0.20\ T^{a}_{(RR)} + 0.16\ T^{a}(HR) + 0.32\ T^{a}_{(PC)} + 0.35\ T_{(VSR)}$$

Where ($\bar{A}$) is the estimated score for the autonomic factor and the letter T following each beta weight indicates the standard scores for the tests shown in parentheses. The superscript (a) on T indicates that these tests are to be reflected. The tests shown in the parentheses of the equation include: Salivary output (SO), pupillary diameter (PD), diastolic blood pressure (DBP), respiration rate (RR), heart rate (HR), palmar conductance (PC), and volar skin resistance (VSR). The measurements of these tests were made as follows:

Measurement of salivary output: - The maximum salivary output was determined for a period of 3 min. The subject was seated and handed over the apparatus for collection of saliva. He was instructed to hold the plastic saliva ejector between his lips and was consistently urged to force as much saliva as possible to the front of his mouth. The 3-min period was timed with the stop watch. The sample was allowed to settle before recording the value. When all the air bubbles had disappeared, the volume was read in the graduated tube. The score was the amount to the nearest 0.1 cc.

Measurement of the pupillary diameter: - Pupil diameter was measured by the technique of Helmholtz as described by Granger (1954). It required a dull black card 13x20cm in size, on which pairs of pinpoint holes were drilled at intervals, so that center-to-center distance between any pair increased in 0.5mm steps. The effective range of measurement with the card was from 2.0 to 8.0mm. Through the pinpoint holes, the subject fixated on object 30 feet or more away. He moved the card slowly up and down until he found a pair of holes which provided tangent images. Before making the measurements the subject was instructed as follows:

"The card permits us to measure the size of the pupil of your eye. With your one hand hold the card over one eye and cover the other eye with your other hand. Now move the card slowly up and down and look through the pairs of holes. Sometimes the holes will overlap like this ⚭, ⚭, sometimes they will touch each other like this ○○, ○○ , and sometimes they will be apart like this ○ ○, ○ ○. Let the experimenter known when you see the pair of circles that just touch each other."

This enabled the subject to see three possible types of patterns in his peripheral vision. If the center-to-center distance of a pair of holes was smaller than the pupillary diameter, the images of the two holes appeared to overlap. If the center-to-center distance between holes was greater than the pupillary diameter the images of the two holes appeared to be two separate circles. If, however, the center-to-center distance between holes was equal to the diameter of the pupil of the eye, the circles of light appeared tangent i.e. just touching each other. On moving the card slowly up and down when the subject had the pair of holes that were seen just tangent the greater part of time, he held the card stationary and the experimenter read the pupillary diameter from the distance between the holes marked on opposite side of the card. The procedure was repeated until three consistent readings were obtained, and the data was their mean value in millimeters.

Determination of blood pressure:- Blood pressure was recorded using a sphygmomanometer (Hunter and Bomford, 1970). A standard adult size (12x23cm) BP Cuff was tied around the upper arm and a stethoscope Chest-piece was fastened over the brachial artery. Arm cuff pressure was recorded during gradual cuff deflation by Korotkoff sound determination of arterial pressure for reading the systolic and diastolic end points.

Determining the rate of respiration:- Respiration was recorded using a thermocouple lead which was held in the breath stream of a subject. The variation in temperature of the inspired and expired air in each breath was sensed by the thermocouple and the electrical changes thus generated were amplified by a low level DC preamplifier and recorded on one of the channels of the polygraph machine. The respiration rate, was analysed by counting the respiratory excursions in every one minute sample, and the recorded datum was the mean of all such samples.

Recording of electrocardiogram for determination of heart rate:- The heart rate was obtained from a continuous recording of conventional EKG Lead II on one of the Channels of the polygraph machine (Grass, Model 7B Quincy, Mass, U.S.A.), EKG was recorded for a period of 15 to 20 min. To obtain the heart rate, the record was analysed by counting the QRS peaks in each one minute sample of the record. The recorded datum was the mean of all such samples.

Measurement of palmar skin conductance: - This test represents the measurement of resistance offered by the palmar surface of the skin to the flow of an applied weak DC current of 40 μA. The resistance was measured by placing one surface electrode each of 2 cm diameter on the middle of each palm using 1% $ZnSO_4$ in Agar - Agar jelly. Measurements were taken in the subjects at supine position, at 1 min intervals until three concordant readings were obtained. The datum for any one subject represented the mean of the three measurements. These values were later converted to conductance unit in mhos.

Volar fore arm skin resistance: - This test measured resistance offered by the skin on the volar area of the fore arm (the inner part of fore arm which rests on the desk when one writes). The apparatus and the recording procedures were identical to those employed for palmar skin conductance described above. The electrodes were placed in the volar surfaces of the fore arm 8-10 cm below the elbows. Samples were obtained in the same manner as for palmar skin conductance.

RESULTS

(A) EXPERIMENTS IN SUMMER

Reactivity in terms of autonomic scores ($\bar{A}$): - Fig. 1 shows the effect of exposure to high T_a of 44° C in Summer, on quantitative $\bar{A}$ estimates. The mean $\bar{A}$ estimate in standard score form, for the subjects at 27° C was 53.37, which is in very close approximity to the population mean. The mean autonomic scores for the same subjects at T_a of 44° C was 41.86, which is significantly lower than that at 27° C ($p < 0.001$). The decrease in the numerical $\bar{A}$ estimate was of the magnitude of one standard deviation (SD) below the population mean and was in the direction of sympathetic (SNS) dominance. (Fig. 1a). Inspection of individual autonomic estimates (Fig. 1b) for the subjects at 27° C revealed that of the thirty subjects, four demonstrated relatively low scores (apparent SNS dominance) eight demonstrated relatively high scores (apparent PNS dominance) and the remaining eighteen subjects demonstrated a balanced autonomic state (score about 50). While at high T_a of 44° C, four subjects showing apparent SNS dominance at 27° C did not show significant change (i.e. deviation was less than one SD in the $\bar{A}$ scores and thus remained SNS predominant, eight subjects showing PNS dominance at 27° C shifted to the mean level and thus acquired balanced autonomic state. Of the eighteen individuals showing mean $\bar{A}$ scores of about 50 at 27° C twelve demonstrated decrease more than one SD and thus shifted towards the SNS direction and the scores of the remaining six subjects remained within the range of autonomic balance score. Thus, the subjects having $\bar{A}$ scores indicating either PNS dominance or balanced autonomic state demonstrated decrease in their scores which, in other words,

(a) MEAN AUTONOMIC SCORE ($\bar{A}$)

Climate (outdoor)	T_a	Mean	SD	T Units	Diff.	t
40.83°C	27°C	86.49	8.08	53.37(53)	1.2 SD towards S	7.41*
29.34RH	44°C	77.59	6.53	41.86(42)		

* Significant at 0.1% level of confidence.

(b)	27° C	44° C			Total
		P	M	S	
	P	0	8	0	8
	M	0	6	12	18
	S	0	0	4	4
	Total	0	14	16	30

S = Sympathetic predominant
P = Parasympathetic predominant
M = Mean (Autonomic balance)

(c)

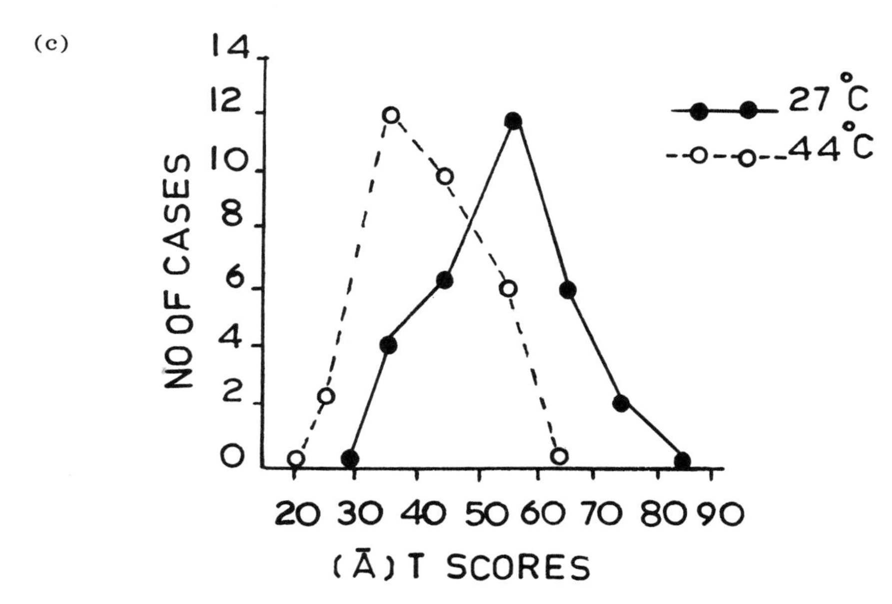

Figure 1: *Effect of Heat exposure on autonomic scores in summer observed on the same individuals (N = 30).*

TABLE 1: *Profile sheet showing the effect of heat exposure (44°C) on autonomic response pattern in summer observed on the same individuals (N=30).*

OUTDOOR CLIMATE	: 40.83°C, 29.34RH	T_a	27°C ✕✕ / 44°C ■■	T Score	= 54 Units / = 42 Units	

Test	S	−3σ −2σ −1σ M +1σ +2σ +3σ 20 30 40 50 60 70 80	P	Normative Group Mean	S.D.
Salivary output	Low		High	3.51	0.99
Diastolic blood pressure	High		Low	68.78	7.50
Heart rate	Fast		Slow	64.12	9.19
Log palmar conductance	High		Low	0.8010	0.17
Volar skin resistance	Low		High	151.35	48.22
Respiration rate	Fast		Slow	18.81	4.65
Pupillary diameter	Large		Small	4.50	0.93
Log palmar resistance change	Small		Large	0.2140	0.11
Sublingual temperature	High		Low	36.93	0.19
Finger temperature	Low		High	34.11	0.66

CLIMATE (OUT DOOR) 40.83°C, 29.34 RH		PATTERN DEFINITIONS SO	DBP[a]	HR[a]	PC[a]	VSR
T_a	27°C	54.36(54) Mean	51.57(52) Mean	50.37(50) Mean	46.82(47) Mean	56.11(56) High (+)
	44°C	49.00(49) Mean	50.50(50) Mean	47.15(47) Mean	42.83(43) High (−)	41.87(42) Low (−)
DIFFERENCE		05 Low (−)	02 High (−)	03 High (−)	04 High (−)	14 Low (−)
	t	3.04	0.49	5.36	7.56	7.09
	p	<.01	NS	<.001	<.001	<.001

(−) indicates a deviation from the normative mean of $\frac{1}{2}\sigma$ or greater in a direction indicative of apparent dominance of Sympathetic Nervous System (SNS) activity. (T score of 45.0 or less).

(+) indicates a deviation from the normative mean of $\frac{1}{2}\sigma$ or greater in a direction indicative of apparent dominance of Parasympathetic Nervous System (PNS) activity. (T score of 55.0 or greater).

a = Reflection of variable; NS = Not significant.

means a shift towards an increase in overall SNS activity or a decrease in PNS activity. The same fact is revealed by the distribution curve (Fig. 1c) which shifted in its entirety towards the SNS end of the scale.

Reactivity in terms of autonomic response patterns:- Table 1 illustrates the profile sheet showing the effect of heat exposure in Summer on autonomic response pattern (ARP). The results of the ARP analysis indicate that the individuals exposed to high T_a in Summer, demonstrated low SO, fast HR, high DBP, high PC and low VSR as compared to their control values. The increase in HR, however was not significant statistically.

(B) EXPERIMENTS IN WINTER

Reactivity in terms of autonomic scores ($\bar{A}$): - Fig. 2 shows the effect of exposure to high T_a of 44° C in Winter, on quantitative ($\bar{A}$) estimates, their means differences from 27° C and distribution curves. The mean $\bar{A}$ estimate in standard score for the subjects at 27° C was 50.59, which is in very close approximity to the population mean. The mean $\bar{A}$ score for the same subjects at an ambient T_a of 44° C was 40.59, and was significantly lower statistically than that obtained at 27° C ($p < 0.001$). The decrease in the numerical $\bar{A}$ estimate was of the magnitude of one SD below the population mean, and was in the direction of SNS dominance (Fig. 2a).

Inspection of individual autonomic estimates (Fig. 2b) for the subjects at 27° C revealed that out of thirty subjects, twelve demonstrated relatively low scores (apparent SNS dominance), twelve demonstrated relatively high scores (apparent PNS dominance) and the remaining six subjects demonstrated a balanced autonomic state (i.e. score about 50). While at T_a of 44° C, twelve subjects having scores indicating apparent SNS dominance at 27° C further decreased their $\bar{A}$ scores and thus continued to show excessive SNS predominance. Whereas of the twelve subjects having scores indicating PNS dominance at 27° C, only two shifted to the mean level and thus acquired autonomic balance state and the remaining ten did not show much change (i.e. deviation was less than half SD) in the $\bar{A}$ scores and thus, remained PNS predominant. While the rest of the six individuals who demonstrated mean $\bar{A}$ scores at 27° C continued to show their scores in the range of autonomically balanced state even at a high T_a of 44° C. Thus, the subjects having $\bar{A}$ scores indicating SNS dominance demonstrated decrease in their scores which, in other words, means a shift towards an increase in overall SNS activity. While, the subjects having $\bar{A}$ scores indicating either PNS dominance or autonomic balance state, demonstrated no significant change in their scores and, thus, continued to demonstrate their previous status which, in other words, means no appreciable change in the SNS or PNS activity. The same fact is revealed by the distribution curves (Fig. 2c).

Reactivity in terms of autonomic response pattern:- Table 2 illustrates the profile sheet showing the effect of heat exposure in winter on ARP. The result of the pattern analysis indicated that the individuals

(a) MEAN AUTONOMIC SCORE ($\bar{A}$)

T_a	Mean±	S.D.	T. Units	Diff.	t.
27° C	84.03±	11.11	50.59(51)	1.0 SD towards S	5.17*
44° C	76.55±	15.81	40.59(41)		

* Significant at 0.1% level of confidence.

(b)

27° C	44° C			Total
	P	M	S	
P	10	2	0	12
M	0	6	0	6
S	0	0	12	12
Total	10	8	12	30

S = Sympathetic Predominant
P = Parasympathetic Predominant
M = Mean (autonomic balance)

(c)

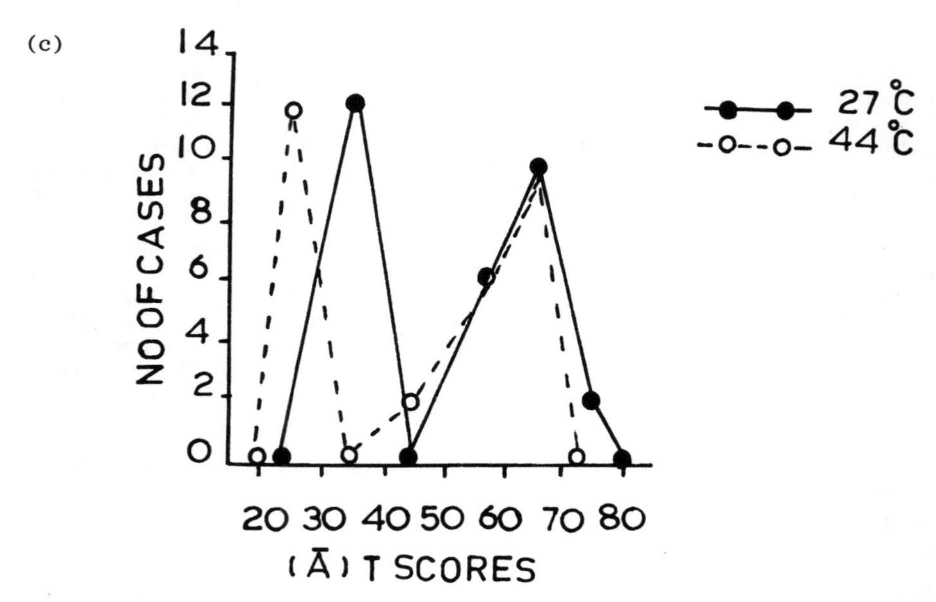

Figure 2: *Effect of heat exposure on autonomic scores in winter observed on the same individuals (N = 30).*

TABLE 2: *Profile sheet showing the effect of heat exposure (44°C) on autonomic response pattern observed on the same individuals in winter (N = 30).*

OUTDOOR CLIMATE	: 21.24°C, 72.35RH	T_a	27°C ✕✕ 44°C ■■	T Score	= 51 Units = 41 Units

Test	S	-3σ 20	-2σ 30	-1σ 40	M 50	+1σ 60	+2σ 70	+3σ 80	P	Normative Mean	Group S.D.
Salivary output	Low								High	3.51	0.99
Diastolic blood pressure	High								Low	68.78	7.50
Heart rate	Fast								Slow	64.12	9.19
Log palmarconductance	High								Low	0.8010	0.17
Volar skin resistance	Low								High	151.35	48.22
Respiration rate	Fast								Slow	18.81	4.65
Pupillary diameter	Large								Small	4.50	0.93
Log palmar resistance change	Small								Large	0.2140	0.11
Sublingual temperature	High								Low	36.93	0.19
Finger temperature	Low								High	34.11	0.66

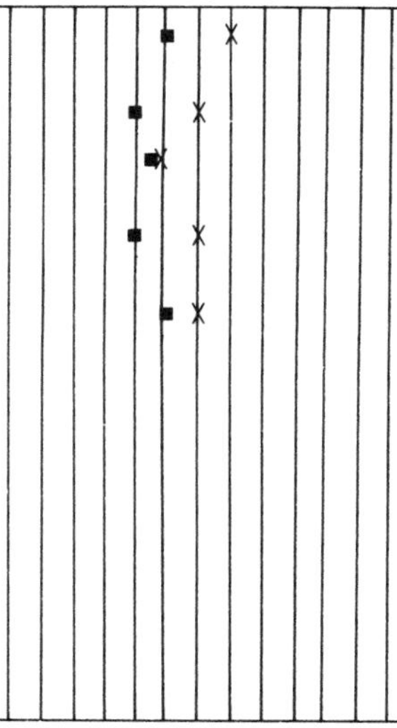

CLIMATE 21.24°C, 72.35 RH		PATTERN DEFINITIONS				
		SO	DBP[a]	HR[a]	PC[a]	VSR
T_a	27°C	54.55(55) Mean	48.55(49) Mean	42.85(43) High (-)	48.49(49) Mean	51.73(52) Mean
	44°C	46.87(47) Mean	41.53(42) High (-)	42.68(43) High (-)	41.62(42) High (-)	45.02(45) Mean
DIFFERENCE		08 Low (-)	07 High (-)	00	07 High (-)	07 Low (-)
	t	5.06	10.49	0.32	4.12	3.85
	p	<.001	<.001	NS	<.001	<.01

(-) indicates a deviation from the normative mean of ½σ or greater in a direction indicative of apparent dominance of Sympathetic Nervous System (SNS) activity. (T score of 45.0 or less).

(+) indicates a deviation from the normative mean of ½σ or greater in a direction indicative of apparent dominance of Parasympathetic Nervous System (PNS) activity. (T score of 55.0 or greater).

a = Reflection of variable. NS = Not significant.

exposed to high T_a in Winter, demonstrated lower SO, higher DBP, higher PC and lower VSR as compared to their control levels at 27° C. The variable HR, however, did not show any significant change statistically.

DISCUSSION

There were significant changes in quantitative estimates of autonomic factor ($\bar{A}$) scores from 27° C to 44° C in two different seasons of the year (Fig. 1 and 2). At a T_a of 27° C the mean $\bar{A}$ scores (in standard score forms) for the two groups were 53.37 and 50.59 in Summer and Winter seasons respectively. Both the group means were thus in a very close approximity to the population mean (i.e. score about 50). The mean $\bar{A}$ scores for the two groups of data were significantly decreased to 41.86 and 40.59 during heat exposure (44° C) in Summer and Winter months respectively. The decrease in the numerical estimate during heat exposure indicates a shift in the direction of SNS dominance which actually means an overactivity of SNS or a decrement in PNS activity. The magnitude of shift in the mean $\bar{A}$ score was about one SD below the population mean in Summer and Winter months.

The decrease in $\bar{A}$ scores at 44° C as compared to 27° C was not uniform in all the subjects and also varied with the seasons. In Winter the change in $\bar{A}$ scores was in general more in the case of those showing SNS dominance and less in case of those who showed PNS dominance or autonomic balance. This is reflected in an increase in the coefficient of variation from 13.22 at 27° C to 20.65 at 44° C in Winter. During the heat exposure studies in Summer, however, the coefficients of variations at 44° C was the same as obtained at thermoneutral T_a of 27° C indicating thereby that individuals in Summer demonstrate shift of almost equal magnitude at higher T_a.

Pattern analysis showing change in the ARP revealed that the individuals exposed to heat stress (44° C) in Summer and Winter demonstrated low SO, fast HR, high BP, high PC and low VSR as compared to their control values obtained at 27° C (Table 1 and 2). It is interesting to note that all the test variables used for pattern analysis were moving in the direction of SNS dominance.

In addition to the main parameters considered for the overall $\bar{A}$ scores as well as for the ARP, there was also a shift towards SNS activity in other functions such as systolic blood pressure and sublingual temperature which showed an increase. Mean weighted skin temperature was also found to be significantly higher in subjects exposed to heat. The observations suggest that under acute heat exposure in addition to changes in heat - responsive-tests as indicated by high palmar and non palmar sweating as measured in terms of electrical conductance of the skin, and cutaneous vasodilation as indicated by high surface temperature, the other physiological test parameters such as HR, BP and SO were also affected. The changes in the physiological variables like SO, HR, BP, palmar and volar skin conductance could be accounted for by the relative hypertonicity of SNS division of ANS, while the high surface temperature at different sites could be due to the generalized vasodilation of the peripheral blood vessels. Because of the practical difficulties involved in obtaining sufficiently large number of individuals with relative SNS OR PNS predominance in a group of unselected population, it was difficult to answer the question of differential reactivity in individuals having different autonomic status. The reason underlying this fact is that the majority of the subjects when studied at thermally neutral temperature of 27° C demonstrate the $\bar{A}$ scores in the range of autonomic balance state. However, the observations available on these subjects throw some light on their reactivity in response to exposure to high T_a in a given season. It is, however, to be noted that no other data are available in the literature which could permit the comparison for this type of work. The details of the thermal influences on reactivity of individuals showing different autonomic status as revealed by the results of the present study are discussed below:

Reactivity of individuals showing SNS dominance to heat exposure: - The individuals showing dominance in the SNS direction at 27° C during Summer and Winter demonstrated decreases in $\bar{A}$ scores on exposure to high T_a of 44° C. On the whole, the decrease in $\bar{A}$ scores was greater in subjects during winter as compared to Summer. This was reflected by the magnitude of deviation which on an average was 1.75 SD in winter as compared to 0.42 SD in Summer (Table 3b). The observation suggests that during Winter (when the acclimatization to cold is maximum in Delhi), the SNS predominant individuals show greater reactivity to a given heat exposure as compared to SNS predominant subjects exposed to heat in Summer season (when the acclimatization to heat is maximum).

Reactivity of individuals showing PNS dominance to heat exposure: - The subjects having $\bar{A}$ scores indicating PNS dominance at 27° C in Summer demonstrated shift in their scores in the range of autonomic balance at high T_a of 44° C; while during winter the individuals demonstrating PNS dominance at 27° C did not show any significant change in the $\bar{A}$ scores and, thus continued to demonstrate PNS dominance even at an air temperature of 44° C. (Table 3). The Winter acclimatized PNS predominant individuals, thus, showed less reactivity to heat exposure, which is reflected in the mean magnitude of deviation which was only 0.54 i.e. about half SD, as compared to the summer acclimatized PNS predominant subjects, where the magnitude of deviation was always greater than 1 SD, being 1.27 SD in Summer.

Reactivity of individuals showing autonomic balance to heat exposure: - Six of the Winter acclimatized subjects having scores indicating balanced autonomic state at 27° C maintained their scores in the autonomic balance state even at 44° C. While twelve out of eighteen (66.6%) subjects in Summer at 27° C, dropped their scores significantly at 44° C and became SNS predominant. The remaining subjects (33.4% in Summer) though demonstrated decrease in their A scores, yet fell into range of mean autonomic balance state. The magnitude of deviation was greater than 1 SD in Summer (being 1.38 SD). While in Winter the shift was only 0.46 SD. The observation suggests that at high air temperature, the reactivity of individuals showing autonomic balance in summer is greater as compared to Winter acclimatized subjects.

TABLE 3a: *Mean Autonomic Factor Scores at 27° C and 44° C in Summer and Winter in subjects Demonstrating Different functional autonomic status.*

	Sympathetic cases		Mean cases		Parasympathetic cases	
T_a	27° C	44° C	27° C	44° C	27° C	44° C
Season						
Summer	36.27(S)	32.07(S)	51.91(M)	39.24(S)	67.11(P)	53.33(M)
Winter	33.90(S)	16.31(S)	55.03(M)	49.59(M)	65.04(P)	62.35(P)

TABLE 3b: *Mean Magnitude of Shift in Terms of Standard Deviation in Autonomic Factor Scores During Heat Exposure (44° C) in Individuals Having Different Functional Autonomic Status.*

Season	Sympathetic cases	Mean cases	Parasympathetic cases
Summer	-0.42(S)	-1.38(S)	-1.27(S)
Winter	-1.75(S)	-0.46(S)	-0.54(S)

The observed decrease in the $\bar{A}$ score in subjects on brief exposure to high T_a in different seasons was in general due to the low SO, fast HR, high DBP, high PC and low VSR. It is to be noted that these five tests appeared in the autonomic regression equation and further more that all these subjects showed deviation in the direction of SNS overactivity.

The above observations suggest that the magnitude of the shift in the $\bar{A}$ scores from the reference mean as measured by using 'NRE' could serve as an index of physiological 'strain' to a physical thermal 'stress' and thus could be used as a functional basis to predict an individual's responsiveness to a particular type of thermal environment.

REFERENCES

ADOLPH, E.F. and MOLNAR, G.W. (1946): Exchanges of heat and tolerances to cold in men exposed to outdoor weather. Am. J. Physiol. 146: 507-537.

AHUJA, M.M.S. and SHARMA, N.J. (1971): Adrenocortical function in relation to seasonal variations in Normal Indians. Ind. J. Med. Res. 59: 1893-1905.

BAJAJ, J.S., GARG, S.K., CHHINA, G.S. and SINGH, B. (1973): Factors influencing plasma cortisol (11-hydroxy corticoids) levels in Rhesus monkeys. Proceedings of 60th (Diamond Jubilee) session of Indian Science Congress. Part III. Abstract p. 717.

BARCROFT, H. and EDHOLM, O.G. (1946): Sympathetic control of blood vessels of human skeletal muscle. Lancet 151: 513-515.

BIRZIZ, L. and HEMINGWAY, A. (1957): Efferent brain discharges during shivering. J. Neurophysiol. 20: 156-161.

CANNON, W.B. (1937): Factors affecting vascular tone. Amer. Heart J., 14: 383-398.

GRANGER, G.W. (1954): Personality and Visual perception. J. Mental Sci. 99: 8-43.

HUNTER, D. and BOMFORD, R.R. (1970): Hutchison's Clinical Methods. Bailliere Tindall & Cassell, London.

LEWIS, T. and PICKERING, G.W. (1931): Vasodilation in the limbs in response to warming the body: with evidence for sympathetic vasodilator nerves in man. Heart 16: 33-51.

LIST, C.F. and PEET, M.M. (1938): Sweat secretion in man II. Anatomic distribution of disturbances in sweating associated with lesions of the sympathetic nervous system. Arch. Neurol. Psychiat. 40: 27-43.

LIST, C.F. and PEET, M.M. (1938): Sweat secretion in man, III. Clinical observations on sweating produced by pilocarpine and mecholyl. Arch. Neurol. Psychiat. 40: 269-290.

RAI, L. (1978): Study of Thermal Influences on Autonomic Functions of the Body, Ph. D. Thesis, A.I.I.M.S., New Delhi.

SARNOFF, S.J. and SIMEONE; F.A. (1947): Vasodilator fibres in the human skin. J. Clin. Invest. 26: 453-459.

THOMAS, P.E. and KORR, I.M. (1952): Significance in areas of low ESR. Fed. Proc. 11: 162.

UVNAS, B. (1954): Sympathetic vasodilator outflow. Physiol. Review. 34: 608-614.

EFFECT OF A RELATIVELY COLD ENVIRONMENT ON THE MEASUREMENT OF THE AUTONOMIC BALANCE IN ADULT MEN

P.L. Ahujarai*, L. Rai**, and S.O.D. Bhatnagar**
(School of Environmental Sciences, Jawaharlal Nehru University, New Delhi)

Abstract: - The effects of a relatively cold environment (outdoor climate: 22.59° C 78.57RH) on reactivity of individuals in terms of quantitative shifts in autonomic balance ($\bar{A}$) and autonomic response patterns (ARP) were studied at an ambient T_a of 20° C on 30 adult healthy men. The data obtained were collated with the study done on the same subjects at a control T_a of 27° C and at an outdoor climate of 32.46° C, 85.26RH. Following exposure to cold environment, there was significant increase in mean $\bar{A}$ score by about 1.7 standard deviation (SD) above the reference mean towards the parasympathetic (PNS) dominance ($P < 0.001$). The pattern analysis for the group revealed that subjects at the low ambient temperature (T_a) demonstrated higher salivary output, larger pupillary diameter, higher systolic and diastolic blood pressures, faster heart rate, lower mean weighted skin temperature and lower finger and toe temperature as compared to the respective control values at 27° C. ANS pattern under cold is thus not of apparent PNS dominance as indicated by high $\bar{A}$ scores but rather a mixed one suggesting thereby hypertonicity of ANS where some components favour SNS activity and some other components manifest PNS activity.

INTRODUCTION

Numerous experimental studies have been conducted to assess the effects of cold stress on cardiovascular (Clark, 1934; Adolph and Molnar, 1946), respiratory, metabolic (Scholander et al., 1950; Scholander, 1959; Hart, 1958), gastrointestinal (Adolph and Molnar, 1946; Louridis, 1970) and several other functions of the body including the renal functions. However, these investigations have been primarily done from the point of view of thermoregulation and no attempt has been made to interpret them as an alteration in the basal autonomic pattern. Therefore, an investigation to determine the basic autonomic response pattern (ARP) of the individuals from different backgrounds

Present addresses:
* Department of Environment, Bikaner House, New Delhi-110011 (India).
** Central Research Institute for Yoga, New Delhi-110001 (India).

of thermal environment and their reactivity to given cold environment might provide clues to their capacity for adjustment in an altered thermal environment. If individuals differ in autonomic functions per se, and if individuals having different functions of autonomic activity react differently in a given thermal stress, the differential reactivity might provide clues to discover individuals who can well withstand the thermal stress due to change in thermal environment. This paper deals with the effects of relatively low ambient temperature (T_a) on reactivity of individuals in terms of autonomic factor scores ($\bar{A}$) and autonomic response patterns. Thirty adult male subjects were investigated in winter months (outdoor climate: 22.59° C and 78.57 RH) in a room whose temperature was kept constant at about 20° C. The results were compared with the findings of the study done on the same subjects under thermally moderate environmental conditions (i.e. during monsoon months: July, August and September) in a room whose temperature was kept constant at about 27° C.

METHODS

Effects of low environmental temperature on reactivity of individuals in terms of autonomic factor ($\bar{A}$) scores and autonomic response patterns (ARP), were studied on 30 adult healthy men. For determining the $\bar{A}$ score in cold environment, a seven test battery i.e. salivary output (SO), pupillary diameter (PD), respiration rate (RR) diastolic blood pressure (DBP), heart rate (HR), palmar conductance (PC), and Volar skin resistance (VSR), was employed. The procedures for measurements of these tests are described elsewhere (Rai, 1978).

The data on the physiological test parameters were entered in the normative regression equation employed for determining quantitative estimates of autonomic balance (imbalance).
The normative equation is:

$$(A) = 0.35\ T_{(SO)} + 0.03\ T^{a}_{(PD)} + 0.28\ T^{a}_{(DBP)} + 0.20\ T^{a}_{(RR)}$$

$$+ 016\ T^{a}_{(HR)} + 0.32\ T^{a}_{(PC)} + 0.33\ T_{(VRS)}.$$

This equation was derived from a factorial study of 110 normal healthy adult males (Rai, 1978). In the equation, ($\bar{A}$) is the estimated score for the autonomic balance and the T following each beta weight indicates the standard scores for the tests shown in parentheses. The superscript (a) on T indicates that these tests are to be reflected. Among the seven tests these are the only ones which bear a negative relationship to the others. A low raw score on these tests, is equal to a high standard score for utilization in the regression equation.

RESULTS

Reactivity in terms of autonomic scores: - Fig. 1 shows the effects of low ambient temperature (20° C) on quantitative estimates of autonomic balance ($\bar{A}$), their mean differences from 27° C and distribution curves. The mean estimate of $\bar{A}$ in standard score for the subjects at 27° C was 48.42, which is in very close approximity to the population mean. The mean $\bar{A}$ score at low environmental temperature for the same subjects was 65.09, significantly higher than that at 27° C ($p < 0.001$). The increase was about 1.7 SD above the population mean towards the direction of PNS dominance (Fig. 1a).

Inspection of individual autonomic estimates (Fig. 1b) for the subjects at 27° C revealed that out of the thirty subjects, four demonstrated relatively high scores (apparent PNS dominance), six demonstrated relatively low scores (apparent SNS dominance) and the remaining twenty subjects demonstrated a balanced autonomic state (i.e. score about 50). At low environmental temperature (or ambient cold), of the six subjects showing apparent SNS dominance at a temperature of 27° C, four acquired $\bar{A}$ score while the remaining two shifted in the direction of PNS dominance, and the four subjects showing PNS dominance at 27° C, did not show any significant change (i.e. deviation was less than half SD) in the $\bar{A}$ scores and thus remained parasympathetic dominant. Of the twenty subjects having scores indicating balanced autonomic activity at 27° C, fourteen demonstrated an increase of more than one SD (some even more than two SD) and thus shifted towards the direction of PNS dominance, while the remaining six subjects maintained their scores in the range of mean level. Thus, the subjects having scores indicating either SNS dominance or $\bar{A}$ score at thermally moderate ambient conditions, show the tendency to have high quantitative autonomic estimates at relatively low environmental temperature (i.e. ambient cold), which in other words means a shift towards an overall increase in PNS activity or a decrement in SNS activity; while the subjects having scores indicating PNS dominance at 27° C tend to maintain their previous scores even at low T_a and thus remain PNS predominant. These facts are revealed by the distribution curve (Fig. 1c) which shows the tendency to shift towards the parasympathetic end of the scale.

Reactivity in terms of autonomic pattern: - Table 1 illustrates the profile sheet showing the effect of low environmental temperature on ARP. The pattern analysis for the group revealed that the subjects in the low T_a demonstrate significantly higher SO, higher DBP, fast HR, lower PC and higher VSR as compared to their control levels. Statistically, however, the DBP did not show a significant change.

DISCUSSION

There were significant changes in quantitative estimates of $\bar{A}$ scores in subjects in moderate and cold environment as compared to the

(a) MEAN AUTONOMIC SCORES

Environment	Mean	SD	T.Units	Diff.	t
32.46° C 85.26RH (27° C) Moderate	82.41	6.16	48.42(48)	1.7SD towards P	5.41*
22.59° C 78.57RH (20° C) cold	94.86	8.97	65.09(65)		

* Significant at 0.1% level of confidence.

(b)

Moderate Environment		Cold Environment 22.59°C, 78.57RH (20°C)			Total
		P	M	S	
32.46° C 85.26RH (27° C)	P	4	0	0	4
	M	14	6	0	20
	S	2	4	0	6
Total		20	10	0	30

S = Sympathetic predominant
P = Parasympathetic predominant
M = Mean (Autonomic balance)

(c)

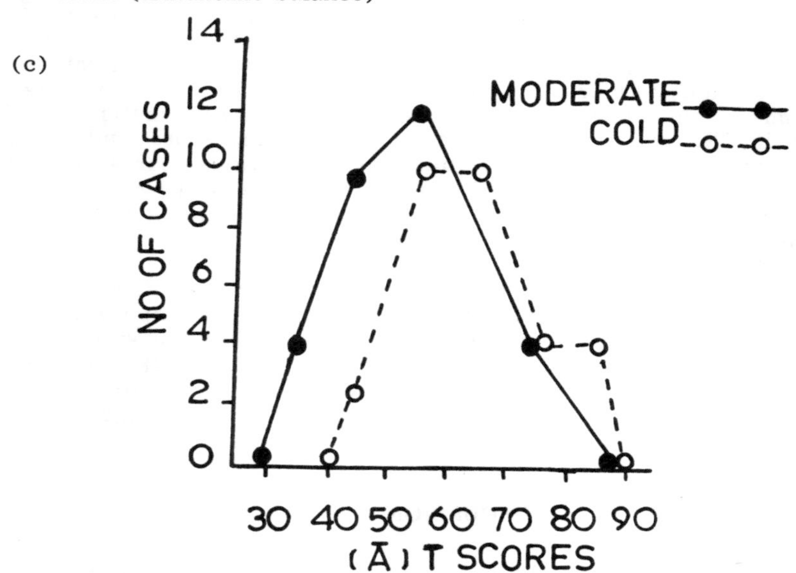

Figure 1: *Effect of cold environment on autonomic scores observed on the same individuals (N = 30).*

TABLE 1: *Profile sheet showing the effect of cold environment on autonomic response pattern observed on the same individuals. (N-30).*

	T_a		T
1. Moderate Environment	27° C	-XX-	= 48 Units
2. Cold Environment	20° C	-■■-	= 65 Units

Test	S	-3σ 20	-2σ 30	-1σ 40	M 50	+1σ 60	+2σ 70	+3σ 80	P	Normative Mean	Group S.D.
Salivary output	Low								High	3.51	0.99
Diastolic blood pressure	High								Low	68.78	7.50
Heart rate	Fast								Slow	64.12	9.19
Log palmar conductance	High								Low	0.8010	0.17
Volar Skin resistance	Low								High	151.35	48.22
Respiration rate	Fast								Slow	18.81	4.65
Pupillary diameter	Large								Small	4.50	0.93
Log palmar resistance change	Small								Large	0.2140	0.11
Sublingual temperature	High								Low	36.93	0.19
Finger temperature	Low								High	34.11	0.66

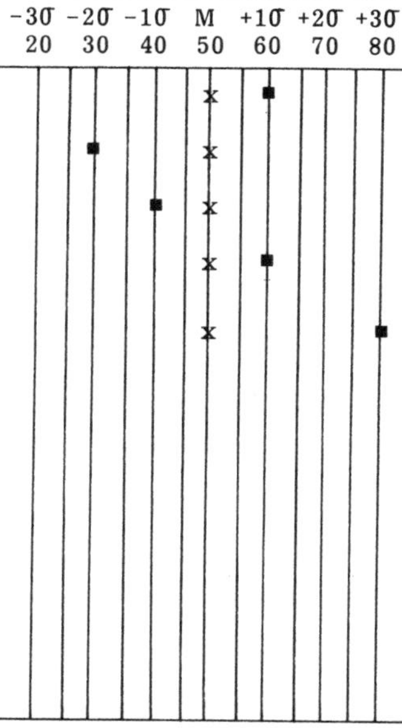

ENVIRONMENT	PATTERN DEFINITIONS				
	SO	DBP[a]	HR[a]	PC[a]	VSR
Moderate Environment 32.46°C 85.26RH(27°C)	50.08(50) Mean	52.41(52) Mean	49.22(49) Mean	49.05(49) Mean	51.17(51) Mean
Cold Environment 22.59°C, 78.57RH(20°C)	60.00(60) High (+)	29.70(30) High (-)	41.88(42) High (-)	60.77(61) Low (+)	78.32(78) High (+)
Difference	10 High (+)	22 High (-)	07 High (-)	12 Low (+)	27 High (+)
t	3.16	1.13	2.56	5.34	5.95
p	<0.01	NS	<0.05	<0.001	<0.001

(-) indicates a deviation from the normative mean of $\frac{1}{2}\sigma$ or greater in a direction indicative of apparent dominance of sympathetic Nervous System (SNS) activity. (T score of 45.0 or less).

(+) indicates a deviation from the normative mean of $\frac{1}{2}\sigma$ or greater in a direction indicative of apparent dominance of Parasympathetic Nervous System (PNS) activity. (T score of 55.0 or greater).

a = Reflection of variable; NS = Not significant.

reference $\bar{A}$ scores. The details of the environmental influence at a controlled T_a of 20° C, on reactivity of individuals having different autonomic status as revealed by the results of present study are discussed below:

Response of sympathetic dominant individuals :- The individuals having low scores indicating apparent SNS dominance in moderate environment demonstrated significant increase in their $\bar{A}$ score (i.e. demonstrated apparent PNS dominance) during exposure to cold environment. The increase on an average was 2.39 SD in the direction of PNS dominance. In another study (Rai, 1978), the increase in $\bar{A}$ scores of a magnitude of 3.41 SD in the direction of PNS dominance was reported in SNS dominant cases in hot environment. The observation suggests that individuals who have low $\bar{A}$ score in hot as well as moderate environments tend to show an increase in $\bar{A}$ scores in cold environment. In other words, the individuals who manifest autonomic imbalance in the direction of SNS dominance in relatively hot and moderate environments, demonstrate PNS predominance in relatively cold environment.

Response of parasympathetic dominant individuals: - The individuals having high $\bar{A}$ scores i.e. manifesting PNS overactivity during thermally moderate environment did not show any significant increase in their $\bar{A}$ scores in cold winter (outdoor climate: 22.59° C, 78.57RH) at 20° C. Of the four subjects available, all the four demonstrate deviation of less than half SD in magnitude in $\bar{A}$ scores. The observation suggests that PNS predominant individuals are more stable and are not affected by changing thermal environment due to changing climates.

Response of individuals showing autonomic balance: - The individuals having scores indicating autonomic balance in thermally moderate conditions tended to demonstrate a shift towards PNS dominance in cold environment. Thus, there was a significant increase in $\bar{A}$ scores (average = 2.34 SD) in subjects during cold. Out of the twenty individuals available, fourteen demonstrated a shift of more than 2 SD and the remaining six showed a deviation of 1 SD or even less than that.

When the above observations are put together, two suggestions emerge from the data. (i) That the measures of $\bar{A}$ scores are positively associated with relatively cold environment i.e. $\bar{A}$ scores tend to increase in ambient cold and (ii) that the individuals having high scores on the autonomic scale i.e. manifesting apparent PNS dominance demonstrate deviation in their scores, in general of less than half SD in magnitude in thermally cold environment. It may, therefore, be concluded that PNS predominant individuals are probably more stable to thermal stress.

The pattern analysis revealed that the data is valid except for higher SO in cold environment which may have been facilitated by the inhibition of sweating as indicated by higher standard score for palmar and volar skin conductances and thus not due to increased PNS activation. All other measures, however, indicate increased SNS activation. The ANS pattern under cold is thus not of apparent PNS dominance as indicated by high $\bar{A}$ scores but rather a mixed one suggesting thereby hypertonicity of ANS where some components favour SNS activity and some other components manifest PNS activity.

REFRENCES

ADOLPH, E.F. and MOLNAR, G.W. (1946): Exchanges of heat and tolerances to cold in men exposed to outdoor weather. Am. J. Physiol. 146: 507-537.

CLARK, G.A. (1934): The Vaso-dilator action of adrenaline, J. Physiol. 80: 429-435.

GRANGER, G.W. (1954): Personality and Visual perception. J. Mental Sci. 99: 8-43.

HART, J.S. (1958): Metabolic alterations during chronic exposure to cold. Fed. Proc. 17: 1045-1046.

LOURIDIS, et al. (1970): Environmental temperature effect on the secretion rate of "Resting" and stimulated Human mixed saliva. J. Dent. Res. 49: 1136-1140.

RAI, L. (1978): Study of Thermal Influences on Autonomic Functions of the Body. Ph. D. Thesis, A.I.I.M.S., New Delhi.

SCHOLANDER, P.F., HOCK, R., WALTERS, V. and IRIVING, L. (1950): Heat regulation in some arctic tropical mammals and birds. Biol. Bull. 99: 237-258.

SCHOLANDER, P.F. (1959): Studies of man exposed to cold. Fed. Proc. 17: 1054-1057.

WENGER, M.A. and CULLEN, T.D. (1972): Studies of autonomic balance in children and adults. In N.S. Greenfield & R.A. Strenbach (Eds.) Handbook of psychophysiology, New York, Holt, Rinehart and Winston, 535-570.

TEMPERATURE ACCLIMATION IN MICE AS SHOWN BY CHANGE IN THE RECTAL TEMPERATURE AND SURVIVAL TIME

L.B. Jha
(School of Environmental Sciences,
Jawaharlal Nehru University, New Delhi - 110 067)

Abstract: - The rise and fall of rectal temperature (Tre) and survival time (min) in mice was determined when exposed to 42° C and -15° C of temperature in the laboratory condition. The results suggest that the rise and fall of temperature in control, previously exposed, and adrenalectomized mice was not significant, whereas results of the experiment of the survival time was significant. It is concluded from the above experiment that the shortterm exposure does not induce physiological changes in the mice to get acclimatized when exposed to 42° C and -15° C, while in the case of survival time it does so.

INTRODUCTION

Acclimatization is the functional compensation over a period of day to weeks in response to a single environmental factor only as in controlled experiments. (Folk, 1969). Acclimatization and survival of rats in cold increase the ability of animals to survive under stressful living conditions. Budd (1962) in his experiments on human subject has demonstrated acclimatization to cold in Antarctica as shown by rectal temperature response to a standard cold stress. Study on the thermal tolerance of fish from a reservoir receiving heated effluent from a nuclear reactor has been done by Holland (1974) who found that in fish obtained from a reservoir having high temperature through out the year, relative to those of natural areas, the thermal tolerance was higher. The present study was conducted to investigate the acclimation in mice to heat and cold stress as shown by rectal temperature change and survival time.

MATERIALS AND METHOD

Adult swiss albino mice weighing between 25-35 g were used for experiments. Induction of thermal stress and recording of rectal temperature and survival time were done in the following way; the high ambient temperature (42° C) was attained by adjusting the control knob of BOD incubator. This gave the desired temperature inside the desiccator which was read with the help of a mercury thermometer fitted into it.

Cold stress (-15° C) at a constant ambient temperature was induced by exposing the mice in a deep-freezer. By adjusting the thermostat control the desired temperature could be obtained. Air temperature was recorded with the help of alcohol thermometer.

Telethermometer (Aplab) using a thermister probe (Yellow Springs, Ohio, U.S.A.) was used for recording the rectal temperature. The probe was inserted into the rectum. The thermister probe was held in position by sticking it to the tail with the help of Johnson tape. The thermister's other end was connected to telethermometer for direct recording of rectal temperature. Calibration of telethermometer was done with the help of mercury thermometer. The survival time of the animal during exposure to heat and cold stress was indicated by the point of cessation of respiration.

Adrenalectomy was done by anaesthetizing the mice with ether. The mice were kept ventral side down and hairs were clipped on both sides of the last rib of the back region. The incision was made on the right side, muscle layer was cut with the tip of scissors and opening was widened (Zarrow et al., 1969). The adrenal gland was seen at the anterior pole of the kidney, which was removed after separating it from fat. After excision of gland, it was kept on a moist paper to make sure that complete gland has been removed carefully. The incision was closed with separate sutures applied to the muscle layer and skin. The left adrenal gland was removed in the same manner. The nostrils of the animal were exposed to an ether cone throughout the operation. The animals were housed at a slightly higher temperature than normally used and 1% NaCl solution was given for drinking purpose.

RESULTS AND DISCUSSION

Results of the experiments relating to the heat acclimation in normal and adrenalectomized mice exposed to heat stress is shown in Figure 1 and 2. The difference in control and pre-exposed mice is not significant. In the case of adrenalectomized mice, there is no significant difference in the rectal temperature rise. It means that the stress was not enough to bring about physiological changes in the animals. In the case of adrenalectomized mice, the non-significance in the rectal temperature change can be explained as follows. Catecholamines play a role in adaptation to thermal stress, i.e., heat and cold stress. It is known that catecholamines in the organisms control non-shivering thermogenesis (Christopherson, 1973). According to this view, the regulation of non-shivering thermogenesis takes place through the noradrenaline released by sympathetic nerve endings. But adaptation is a long-term and genetic process, so this mechanism is not possible for short time exposure, i.e., the gap between 1st and 2nd exposure. This may be the reason for non-significant difference in the rectal temperature in the experimental animals in our experiment. Heat

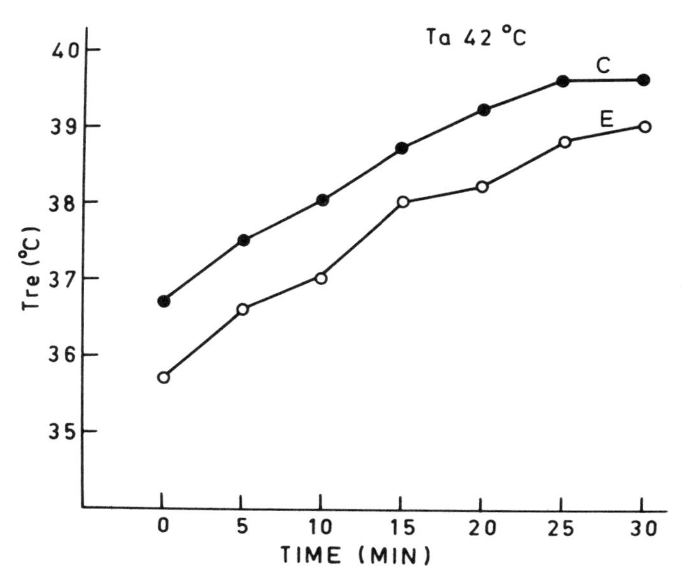

Figure 1: *Showing rectal temperature change in mice previously exposed to 42° C.*
C - Control; E - Experimental (Pre-exposed).

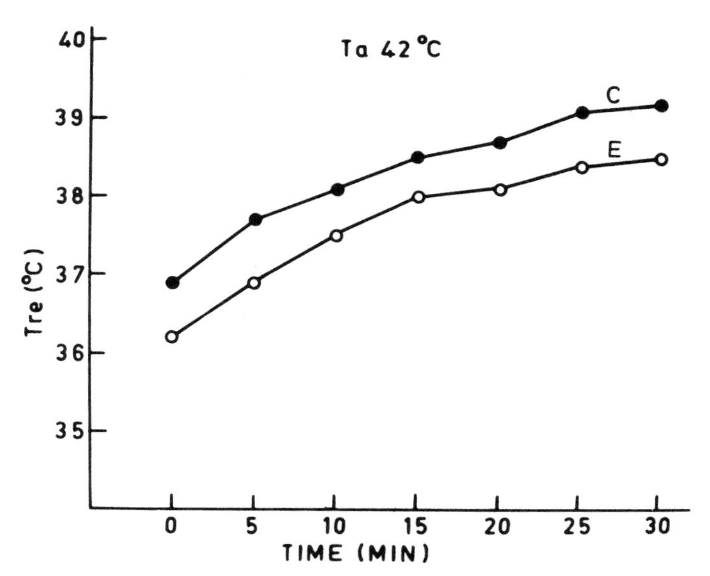

Figure 2: *Showing rectal temperature change in adrenalectomized and normal mice previously exposed to 42° C.*
C - Control; E - Experimental (adrenalectomized).

stress also activates the adrenal cortex. Glucocorticoids and aldosterone are released in the process. But no direct evidence is available in favour of the role of glucocorticoids in thermal adaptation. The role of aldosterone in relation to heat adaptation has been confirmed. It stimulates the reabsorption of NaCl and so its concentration is lowered in the sweat and urine. This is regarded as most characteristic modification for heat adaptation.

Fig. 3 shows the change in rectal temperature in mice, in response to cold stress, which was exposed to the same temperature for half an hour one week earlier. The difference in rectal temperature is not significant. Similar results were found in the case of adrenalectomized mice as shown in Fig. 4. The effect of adrenalectomy on the physiological response of the animals exposed to cold is controversial. It also did not show significant difference. The interaction between stress and physiological effect was prevented by hairs which did not allow the heat from the body to dessipate rapidly as has been found in shaved rats and man. The other reasons for the rectal temperature difference not being significant might be species difference and less body surface area.

In the case of survival time there is significant difference in control and pre-exposed animals which indicates that certain amounts of acclimation has taken place, which prolonged the survival of animals at high environmental temperature. The results are shown in Fig. 5. It is known that thermal stress activates the adrenal cortex thereby releasing glucocorticoids and slight aldosterone. The aldosterone's role has been confirmed; it lowers the concentration of sodium chloride in the sweat and urine. This mechanism might have helped the animals to have enhanced survival time. There is a significant difference in the survival time ($P < 0.001$) between control and experimental animals.

As regards to the survival time in cold exposed animals, control rats survived longer. This might be due to more heat production in control mice. Catecholamine secretion by adrenals helps in the formation of brown fat. It is possible that the release of catecholamine takes place thereby causing non-shivering thermogenesis, and due to increased heat production, the animals survive for longer duration; but the experimental animals die before controls, suggesting that after sometime, the control mechanism fails in the pre-exposed mice which could not sustain the stress adaptation.

CONCLUSION

It can be concluded that the time and frequency of exposure to thermal stress in experimental mice was inadequate and hence the acclimation did not take place, as shown by rectal temperature response. But in the case of survival time in heat and cold stress, the acclimation took place in mice exposed to heat stress due to possible mechanisms discussed above. But in the case of cold exposure, it seems that after certain time the mechanism of thermoregulation fails and animals die prior to the control mice which were exposed previously for a short time.

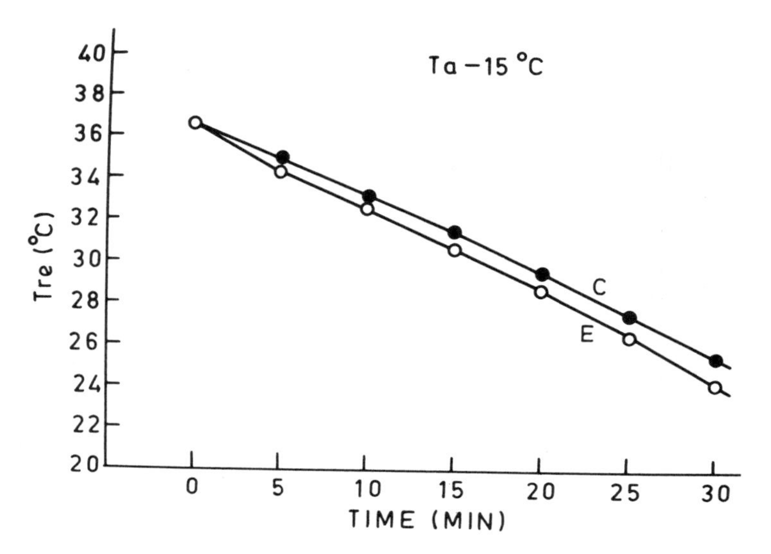

Figure 3: *Shows change in rectal temperature of mice exposed to -15° C.*
C - Control: E - Experimental (Pre-exposed).

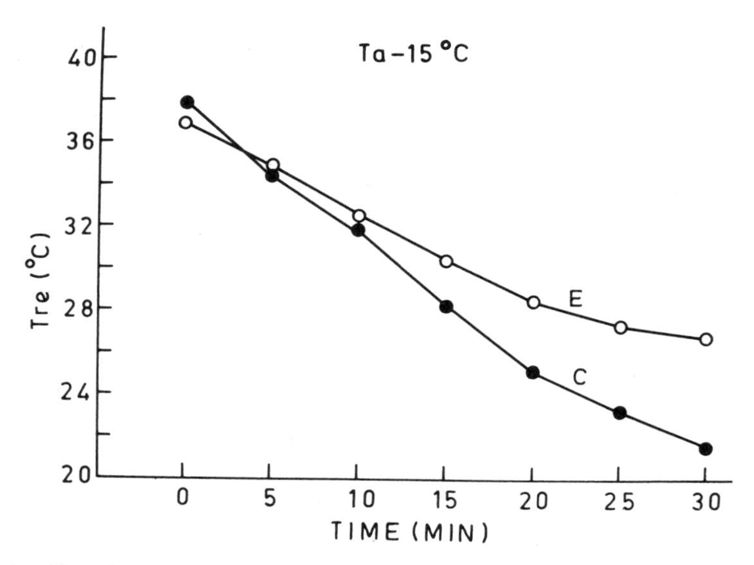

Figure 4: *Showing rectal temperature change in adrenalectomized mice previously exposed (-15° C).*
C - Control: E - Experimental (adrenalectomized).

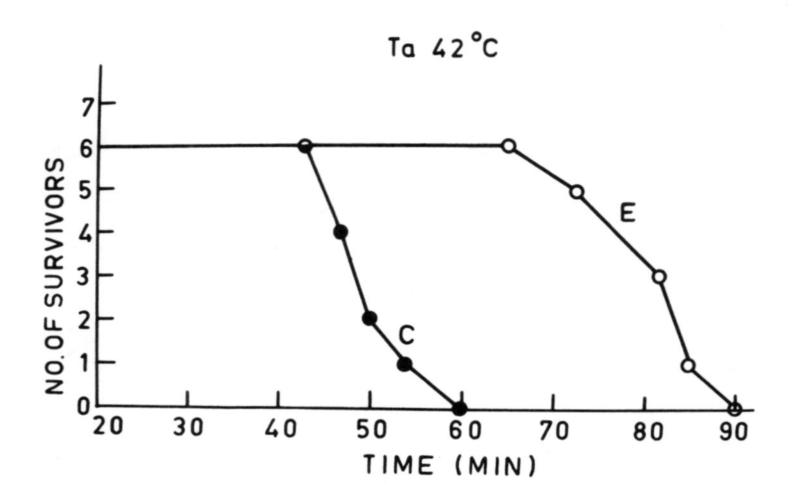

Figure 5: *Showing a change in number and survival time in control (C) and previously heat exposed (E) mice.*

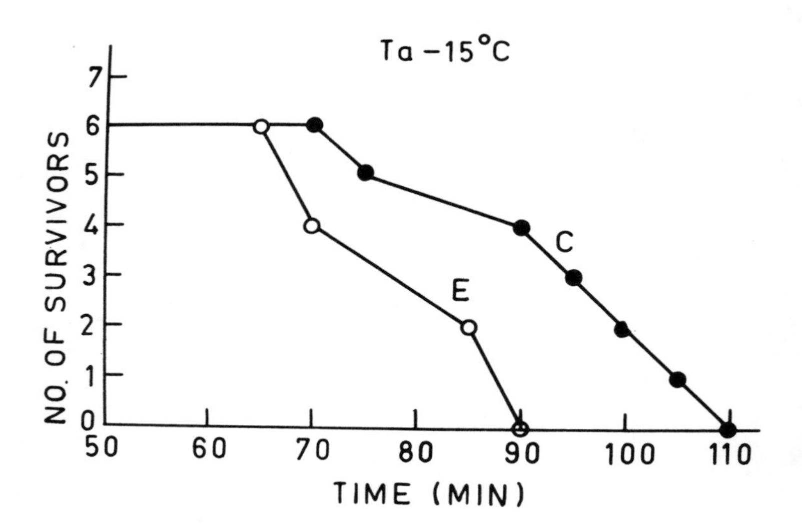

Figure 6: *Showing change in survival time in adrenalectomized mice (E) which was previously exposed to -15° C, as compared to controls (C).*

REFERENCES

BHATIA, B. and SUBRAMANIAN, R. (1974): Changes in resistance to hypoxia. Indian J. Med. Res., 62: 1928-1936.

BUDD, G.M. (1962): Acclimatization to cold in Antarctica as shown by rectal temperature responses to a standard cold stress. Nature, 193: 886.

HALES, J.R.S. (1974): Physiological responses to heat. In: Environmental Physiology, Robertshaw (Ed.) Butterworths, 107-162.

HOLLAND; W.E. et al. (1974): Thermal tolerance of fish from a reservoir receiving heated effluent from a nuclear reactor. Physiological Zoology, 47: 110-118.

FOLK, G.E. (1969): In: Introduction to environmental physiology. Lea & Febiger, Philadelphia.

SELLERS, E.A. et al. (1951): Acclimatization and survival of rats in the cold: effects of clipping, of adrenalectomy and of thyroidectomy. Am. J. Physiol., 165: 481-485.

ROBINSON, S. and WEIGHMAN, D.L. (1974): Heat and humidity. In: Environmental Physiology, Slonim, N.B. (Ed.), The C.V. Mosby Company, Saint Louis, 84-112.

WEBSTER, A.J.F. et al. (1974): Physiological effects of cold exposure. In: Environmental Physiology, Robertshaw, D. (Ed.) Butterworths, University Park Press, 37-70.

STUDY OF SOME HEMATOLOGICAL TRAITS AMONG HIGH ALTITUDE HUMAN POPULATIONS

A.K. Kapoor and Satwanti Kapoor
(Department of Anthropology,
University of Delhi, Delhi - 110 007)

Abstract: - In the present investigation some hematological traits including hemoglobin level, RBC count, WBC count and MCH, were studied among Bhotias, settled and migrant, of District Pithoragarh, Uttar Pradesh. Although an increase in the hematological traits studied has been observed with an increase in the altitude, the values of various traits are not significantly higher as compared to the values of sea-level populations.

INTRODUCTION

The primary environmental stress found at high altitude is 'hypoxia', the lowered partial pressure of oxygen. Among the most important physiological processes in the adaptation of man to his environment at high altitude are those which serve to minimize tissue hypoxia.

Many studies among native men of high altitude have concluded that the rise in the number of erythrocytes and the hemoglobin content are the major adaptive mechanism (Hurtado, 1964; Lenfant and Sullivan, 1971; Garrutto, 1973). All these changes are directed mainly to increase the availability and pressure of oxygen at tissue level.

Morpurgo et al. (1976) in their study on Sherpas of Nepal residing permanently at an altitude of 4,000 m above sea level reported a negligible rise in the hemoglobin level and red cell count. Morpurgo et al. (1976) hypothesized that in Sherpas adaptation does not appear to be brought about by increasing the hemoglobin quantity and red blood cells. It can be either by lowering the hemoglobin-oxygen affinity (improving the release of oxygen to tissues) or by increasing the affinity (improving the oxygenation of blood).

Although a large number of studies have been conducted on hematological traits with regards to age, sex, race, geographic and economic conditions, the studies of these traits on migrant populations living at high altitude are scanty.

The present study aims to present data on some of the hematological traits among the Bhotias (settled and migrant) of District Pithoragarh, Uttar Pradesh, India.

MATERIALS AND METHODS

The present data was collected on Bhotia males. The border regions of Kumaon and Garhwal Himalayas in India are inhabited by a number of semi-nomadic ethnic groups who, when referred to collectively, are known by the generic term "Bhotias". Bhotia is an exclusive community, different both from the Hindus and Tibetans. They are predominantly mongoloid in racial characteristics and speak Tibeto-Burman branch of dialect. The Bhotias of Kumaon and Garhwal are mainly concentrated in the Pithoragarh, Chamoli and Uttarkashi District of Uttar Pradesh State. They are ethnically divided into five endogamous groups namely, Johari, Rang, Jaad, Tolcha and Marchha Bhotias. Rang and Johari Bhotias of the district Pithoragarh have been studied now. Due to environmental and historical factors Bhotias have been leading a semi-nomadic and transhuman life which was till recently associated with the "Trans-Himalayan Trade" with Tibet. This trade was, however, stopped subsequent to the Sino-Indian conflict of 1962. Rang Bhotias are settled in three valleys, viz.- Darma, Byans and Chaudans. The Bhotias of these valleys are collectively known as Rang Bhotias; Johari Bhotias are settled in Johar valley. Rang Bhotias from Darma and Byans valleys migrate to lower altitude (920 m) in winter (November to March) and again move to their respective valleys at higher altitudes (3,530 m; 3,736 m respectively) in summer (April to October). The Rang Bhotias of the Chaudans valley are settled at 1,970 m to 2,550 m. Johari Bhotias are also settled at 760 m to 2,120 m. The former two groups of Rang Bhotias are referred to as migrant Bhotias and the latter two as settled Bhotias in the present text. All the Bhotias studied were native to the place and had been living there since birth. The general health status of all subjects was assessed by outward physical appearance, questions about current and past illness. Individuals with the history of any chronic diseases were excluded from the sample.

All blood samples were collected using the microtechnique of finger puncture (capillary blood). Deep punctures were made to ensure a free flow of blood with as little admixture of tissue fluid as possible.

Hemoglobinometry was performed by Sahli's method based on the conversion of hemoglobin pigment into acid hematin and then matching it against the standard colour tubes using 1/10 HCl as diluting medium. The apparatus used in the field was standardized with a standard Van Slyke and Neill (1924) method based on the oxygen combining power of hemoglobin pigment. Red cell counting was done on the Neubar Chamber by observing it under the microscope. The counting was done in the 80 small squares. The counting was repeated a second time for each sample and the sample was rejected if the difference between the first and second counting was more than 10%.

The white cell count was done on the same Neubar chamber. The counting was done under the microscope using Turk solution as the diluting medium. The corpuscles present in the four corner squares were counted.

The Mean Corpuscular Hemoglobin (MCH) which gives the amount of hemoglobin present in a single corpuscle in absolute terms was also determined for each subject.

TABLE 1: *Hematological findings Among the Bhotias of Himalayas*

Population	Age (yrs)	No. Tested	Altitude (m)	Mode of Settlement	Hb (g/100 ml) Mean±S.D.	RBC ($10^6/mm^3$) Mean±S.D.	WBC ($10^3/mm^3$) Mean±S.D.	MCH (pg)
Johari Bhotias	15-48	135	760-2,120	Settled	13.5±1.01	4.25± 0.21	5.62±1.12	26.15±1.78
Byans Valley	16-53	82	920-3,740	Migrant	14.6±0.78	5.21±10.32	6.31±1.45	28.92±2.11
Rang Bhotias								
Chaudans Valley	15-49	42	1,970-2,545	Settled	13.6±1.12	4.53± 0.58	5.34±1.16	26.25±2.74
Darma Valley	15-51	80	920-3,530	Migrant	15.6±1.31	5.62± 1.61	6.67±1.71	29.17±3.01

TABLE 2: *Hematological Findings Among Adult Men Resident at various Altitudes (Mean values).*

References	Altitude	Sample size	RBC ($10^6/mm^3$)	Hb (g/100 ml)	MCH (pg)	WBC ($10^3/mm^3$)
Hurtado (1932)	4,500 m	25-132	6.70	15.9	24.40	5.20
Salguero-Silva (1971)	3,700 m	-	5.30	16.5	31.10	-
Okin et al. (1966)	3,100 m	65	5.40	17.7	32.70	-
Anderson & Mugraga (1936)	1,600 m	40	5.40	16.5	30.50	-
Okin et al. (1966)	1,600 m	95	5.30	16.6	31.00	-
Singh et al. (1980)	3,030 m	130	4.98	14.6	30.00	6.55
Present Study	760-2,120 m	135	4.25	13.5	26.15	5.62
Present Study	920-3,740 m	82	5.21	14.6	28.92	6.31
Present Study	1,970-2,545 m	42	4.53	13.6	26.25	5.34
Present Study	920-3,530 m	80	5.62	15.6	29.17	6.67
Napier & Das (1935)	Sea level	50	5.36	14.7	-	6.50
Albritton (1952)	Sea level (U.S.)	-	5.40	15.8	29.00	7.40
Williams et al. (1972)	Sea level (U.S.)	186	5.10	15.5	30.20	7.20

RESULTS AND DISCUSSION

Table 1 presents the means and standard deviations of various hematological traits. The minimum average hemoglobin level (13.6 g/100 ml) is shown by the Johari (settled) Bhotias and the maximum (15.6 g/100 ml) by Rang Bhotias (migrant) of Darma Valley. On the whole, the maximum values for all the hematological traits has been displayed by the Rang Bhotias of Darma valley (Hb = 15.6 g/100 ml, RBC = 5.62 $10^6/mm^3$, WBC = 6.67 $10^3/mm^3$ and MCH = 29.17 pg).

The values for different hematological traits is not unusually high as compared to various populations studied at sea level (Table 2). Moreover these values are quite low as compared to many high altitude populations studied (Table 2).

A review of the previous findings reports an increase in the various hematological traits as a compensatory process against the hypoxic stress at high altitude. Hurtado (1932), however, discussed the problem of not reporting an increase in the hematological standards among native Indians of Peru (4,540 m). He concluded that, although an increase is a common finding among highlanders, a normal count is also not incompatible. Findings of Morpurgo et al. (1976) on Sherpas (4,000 m) also support that increase in hemoglobin and erythrocytes is not an essential phenomenon for living at higher altitudes. According to him the Sherpas are well adapted to high altitude life even at the normal amount of hemoglobin and RBC number, and has put forward that their adaptation to hypoxic stress is probably by increasing the affinity for oxygen and facilitation of oxygen dissociation and oxygen extraction at tissue level and thereby improving the supply of oxygen to the tissues. Singh et al. (1980) reported similar hematological findings among the Bodhs of Lahaul and Spiti (3,240 m).

The results of the present study also support the findings of Hurtado (1932), Morpurgo et al. (1976) and Singh et al. (1980) that, although an increase in the various hematological standards is known as one of the primary adaptive mechanisms against hypoxia, a normal amount is also not incompatible among high altitude populations. It can be concluded that the Bhotias are well adapted to life at their respective altitude at the normal amount of hemoglobin and red cell number.

REFERENCES

ALBRITTON, E.C. (1952): Standard values in blood. Saunders, Philadelphia.

ANDERSON, M.E. and MUGRAGE, E.R. (1936): Red blood cell values for normal men and women. Arch. Int. Med. 58: 136-146.

GARRUTO, R.M. (1973): Polycythemia as an adaptive response to hypoxic stress. Ph. D. Thesis, Pennsylvania State University, Pennsylvania.

HURTADO, A. (1932): Studies at high altitude - Blood observations on Indian Natives of Peruvian Andes. Am. J. Physiol. 100: 487-504.

HURTADO, A. (1964): Animals in High Altitudes. Resident Man, pp. 843-860. In D.B. Dill, E.F. Adolph and C.G. Wilber (eds.). Handbook of Physiology, Sec. 4, Adaptation to the environment. American Physiological Society, Washington, D.C.

LENFANT, C. and SULLIVAN, K. (1971): Adaptation to High Altitude. New Eng. J. Med. 284: 1298-1309.

MORPURGO, G., ARESE, P. and BOSIA, A. (1976): Sherpas living permanently at high altitude: A new pattern of adaptation. Proc. Nat. Acad. Sci. U.S.A. 73: 747-751.

NAPIER, L.E. and DAS GUPTA, C.R. (1935): Normal standards for a Bengal Town population. Ind. J. Med. Res. 23: 305-309.

OKIN, J.T., TREGER, A., OVERY, H.R., WEIL, J.V., and GROVER, R.F. (1966): Hematological response to medium altitude. Rocky Mt. Med. J. 63: 44-47.

SALGUERO-SILVA, H. (1971): Indices hematóicos normales en La Paz. Unpublished manuscript, Instituto Nacional de Salud Occupacional, La Paz, Bolivia.

SINGH, I.P., BHASIN, M.K. and SINGH, K. (1980): A few haematological responses at high altitude: The Bodhs of Lahaul and Spiti, pp. 187-193. In: Man and his Environment. Indra P. Singh and S.C. Tiwari (eds.) Concept Publishing Co. New Delhi.

VAN SLYKE, D.D. and NEILL, J.M. (1924): The determination of gases in blood and other solution by vacuum extraction and manometric measurements. J. Biol. Chem. 61: 523-573.

WILLIAMS, W.J., BEUTLER, E., ERSLEV, A.J. and RUNDLES, R.W. (1972): Hematology, McGraw-Hill, New York.

CHROMOSOMAL ABERRATIONS IN HIGH ALTITUDE NATIVES AND IN LOW LANDERS INDUCTED TO HIGH ALTITUDES

H. Bharadwaj, T. Zachariah, S. Kishnani, S.N. Pramanik and I.P. Singh
(Defence Institute of Physiology and Allied Sciences, Delhi Cantt-110010 (India))

Abstract: - The effect of exposure to high altitudes on the nature and frequency of chromosomal aberrations was studied on natives as well as on other soldiers from the plains. Photomicrographs of cultured cells revealed that the frequency of aneuploid and tetraploid cells was considerably elevated at high altitude in both the groups. Losses from the G group were most pronounced. Monosomies in C, D and E groups were fairly common. A, B and F groups showed small chromosome loss. Trisomies were mainly observed in D, E and G groups and were not observed in A, B and F groups. Statistical analysis indicates that chromosome loss at high altitude is nonrandom and could be due to mitotic faults.

INTRODUCTION

The last decade has witnessed a spurt in the research on human chromosome pathology and mutagenesis. Inspite of such phenomenal growth in human cytogenetics, little work has been done concerning the effect of oxygen deficiency on the chromosomal structure and function in human cells undergoing mitosis. Teratogenic effects of anoxia were reported by Ingalls et al. (1950). They (1972) subsequently demonstrated that simulated high altitude hypoxia equivalent to 9,100 m to 10,600 m could be a chromosomal mutagen and could cause triploidy and tetraploidy in the hamster embryo. These findings prompted the authors to investigate whether high altitude hypoxia of a longer duration, though of a lesser magnitude, could be effective in producing abnormal chromosomes in man as well. The picture of chromosomal aberration in the high altitude natives vis a vis inductees from the plains could provide more insight into the cytogenetic situation. The present study aims to fill some gaps in this aspect.

MATERIALS AND METHODS

Whole blood micro cultures were obtained on 30 healthy Indian infantry soldiers aged 21-30 yrs. On the basis of selection criteria described below they were grouped in two categories.

Group I consisted of 15 Ladakhi soldiers born and brought up at altitudes ranging from 3,048 m to 3,962 m. They had neither visited the plains before nor had they stayed at altitudes higher than specified above.

Group II included 15 non Ladakhi men from the plains who were exposed to altitudes over 4,572 m for six months. Their blood cultures were obtained at a field hospital situated at 3,505 m above sea level. This group also included 5 men from the Indo-Tibetan Border Police who had been exposed to altitudes ranging between 5,047 m to 5,352 m for six months. Blood cultures were immediately prepared on their descent to the field hospital.

Blood Culture Technique:- Standard procedures for making micro blood cultures were followed. Chromosome 4 medium of GIBCO along with its diluent was used to avoid the necessity of a completely sterile chamber. Four drops of heparinised human venous blood drawn from the antecubital vein were inoculated into 5 ml chromosome medium.

This inoculation was then incubated at 37° C for 72 hrs. Colchicine was then added to the culture to attain a final concentration of 0.2 µg/ml. This culture was further incubated for 2 hrs before harvesting. Following hypotonic treatment with 0.075 M KCl at 37° C for 15 min, cells were fixed with 3 changes in a fixative of methyl alcohol and acetic acid in 3:1 ratio. Two drops of cell suspension were dropped on to a clean glass slide, and proper spreading was accomplished by blowing vigorously at right angles to the surface of the slide and passing the slide through a spirit lamp flame.

The slides were washed thoroughly and stained for 20 min in 10% Giemsa-buffer solution (Sorenson, pH 6.8). The stained slides were rinsed briefly in water, dried, passed through Xylene and mounted in DPX mountant. Photographs of well spread metaphases were taken using bright field optics and green interference band filter. (Leitz No S 542-19). A slow speed film was used for better resolution.

EVALUATION OF ANEUPLOID METAPHASES

For evaluating numerical aberrations of chromosomes 10 well spread metaphases were selected at random for each individual in all the study groups. The combination of Leitz oil immersion 100 x apochromatic objective lens and 10 x periplan eye piece provided adequate magnification for the microphotographs. Two copies of photo enlargements were obtained for each of the photo micrographs. Chromosomal counts were made at the back of the photograph using an X-ray film viewer. These were counter checked by crossing the count marks. Metaphases showing abnormal chromosome counts were karyo-typed. The other copy was used for this purpose. Each karyogram was thoroughly inspected for any monosomy, or trisomy, for chromosomal gaps and breaks, polycentric or ring chromosomes. For scoring chromatid or chromosome breaks the method of Mueller (1971) was adopted. Other photo micrographs showing normal chromosome counts were also inspected for the above mentioned anomalies.

ENUMERATION OF TETRAPLOID METAPHASES

For determining the frequency of tetraploid metaphases in the blood cultures, 10 men were selected in each group whose blood cultures had plenty of metaphases. A slide was then randomly selected from each of these cultures and all cells in mitosis were completely enumerated. Any metaphase showing more than 23 chromosomes was included in the count.

Care was exercised in counting tetraploid metaphases so that two overlapping normal metaphases were not included in the count. In the majority of such cases, size, staining characteristics and chromatin condensation distinguished chromosomes from two different metaphases.

A magnification of 250 times was used for general scanning. Tetraploid metaphases were confirmed at 400 x magnification. Photo micrographs of some of the well spread tetraploid metaphases were also taken.

RESULTS

The frequency of aneuploid cells was significantly elevated at high altitude. Amongst the natives at their own habitat, this was 31.4% of the metaphases examined. In other soldiers stationed at high altitudes this was 27.4%. In comparison to these, the frequency of aneuploid metaphases in normal Indian subjects residing at the plains was 3% (Ghosh and Singh, 1974). In other studies conducted in the plains the aneuploid frequency varies from 1.9% (Mueller, 1971) to 9.8% (Jacobs et al., 1964). In the high altitude natives chromatid breaks and gaps were common but chromosome aberrations were relatively few. Total aberration frequency was 4.9% of the metaphases examined. This compared well with the frequency (4.3%) obtained by Gosh and Singh (1974) on 200 Indians in the plains. Littlefield and Goh (1973) have reported a frequency of 5.6% in their 10 control men.

An analysis of aneuploid metaphases of Group I with chromosomal constitution varying from 22 to 53 chromosomes showed 227 chromosome losses. The frequency of loss in each chromosome group has been shown in Table 1. Such assessment has been made purely on centromere position and its overall size in the metaphase.

The smallest chromosome in the G group was the most frequently missing chromosome. This was followed by the smallest chromosome in group E, i.e. 18. Though C group shows the highest loss percentage this may be due to its larger numerical size. In the C group the smaller 10 and 11 chromosomes were conspicuous absentees. The sixth and X chromosome were least missing. Chromosomes of groups A and F followed. Table 1 also gives the frequency of chromosome loss for the soldiers in group II. There is a fair degree of similarity in the pattern of chromosome loss in the two populations at high altitudes, except that the high altitude natives show greater loss in the E group chromosomes.

Among the natives trisomies were most frequent in the D group. Of

TABLE 1: *A Comparison of Expected Chromosome Loss with that Observed Experimentally: Loss Weighted to Chromosome Size.*

Group	Loss Probability for the Group	Population Group: High Altitude Natives Total Chromosome Loss: 227			Population Group: Soldiers from Plains exposed to High Altitudes Total Chromosome Loss: 110		
		Expected Random Chromosome Loss in the Group	Observed Chromosome Loss	Z & P	Expected Random Chromosome Loss in the Group	Observed Chromosome Loss	Z & P
A	0.044	10.00	13	0.029 NS	4.80	5	0.050 NS
B	0.0367	8.30	7	0.462 NS	4.03	2	1.056 NS
C	0.171	38.70	73	6.080 Significant ($P < 0.001$)	18.80	42	5.860 Significant ($P < 0.001$)
D	0.095	21.00	19	0.474 NS	10.46	12	0.500 NS
E	0.107	24.00	45	4.600 Significant ($P < 0.001$)	11.80	14	0.690 NS
F	0.087	19.70	16	0.895 NS	9.50	9	0.148 NS
G	0.140	31.80	54	4.263 Significant ($P < 0.001$)	15.40	27	3.182 Significant ($P < 0.02$)

NS = Not Significant

all trisomies 40.4% involved D group chromosomes, 26.3% E group, 19% C group and 14.3% G group chromosomes. In the soldiers acclimatized to high altitudes, trisomies were detected in D and E groups only. The frequency of tetraploid metaphases in the natives was 0.72. The soldiers from the plains at high altitude also showed the same value. In the plains dwelling population of Delhi (260 m) and elsewhere such frequency ranged from 0.20 to 0.33 (Ghosh and Singh, 1974; Higurasi and Conen, 1971; Knautila et al., 1976). From the available data it is clear that almost a threefold increase in the frequency of tetraploid cells occurs as a result of altitude stress in normal individuals. Natives are no exceptions.

Since chromosomal loss from a metaphase could be due to the preparation technique too, the probability of accidental loss for each chromosome was considered to be 1/46 in the absence of any size criterion. In order to minimize the chromosome size effect in such loss, the chromosome group F was chosen as the reference group. This was done because it was, size wise, the smallest group showing smallest chromosome loss in both sets of data. For giving size weightage to the loss probability, the proportion of the average length of the chromosomes in the specified groups to the average length of the F group chromosomes was considered.

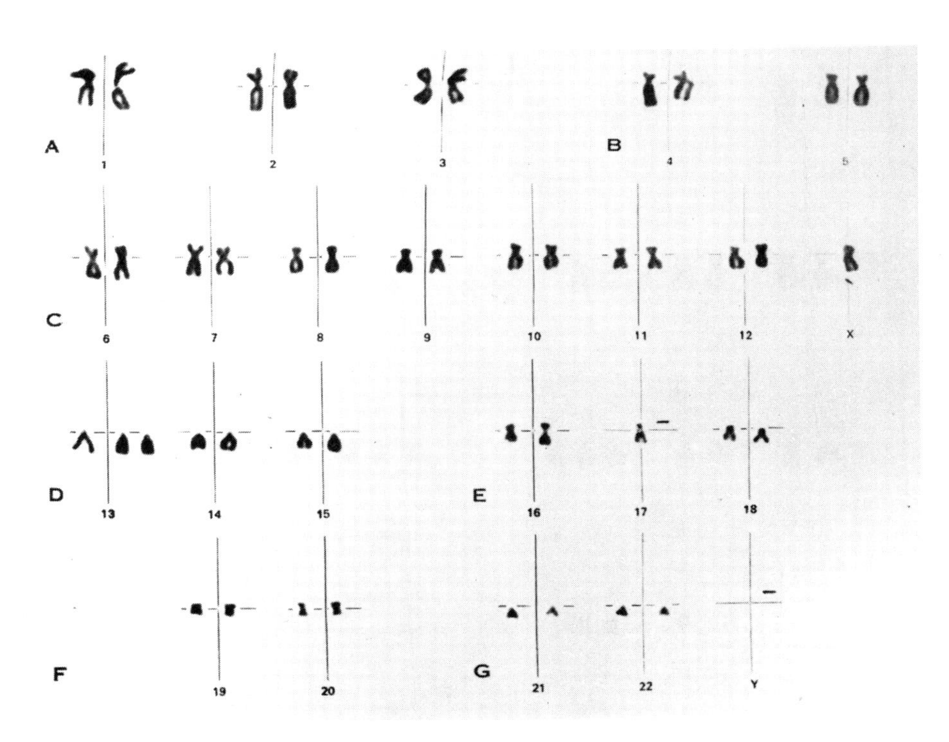

Figure 1: *Trisomy 13, monosomy 17 and missing Y chromosome.*

The expected loss probability for each group was worked out as follows:

Number of chromosomes constituting the group multiplied by

$$\frac{1}{46} \times \left(\frac{\text{Average length of the chromosomes in group (F)}}{\text{Average length of the specified group of chromosomes}}\right)$$

According to this scheme larger chromosomes than that of group F would have smaller loss probability. The data of Lubs et al. (1970) was utilised for calculating different chromosomal length proportions. A comparison of observed and expected loss proportions has been made in the table. Standard normal variate concept was utilised for comparing the expected and observed loss proportions. A value of Z exceeding 1.645 (single tail test) is significant at $P < 0.05$. Inspite of this allowance the observed chromosome loss in C, E and G groups was significantly in excess of the size weighted expected random loss. The results of such analysis indicate that excess chromosome loss could be explained on the basis of mitotic errors rather than errors of preparation technique.

Figures 1 to 4 illustrate a variety of structural abnormalities encountered in the natives and other soldiers stationed at high altitude.

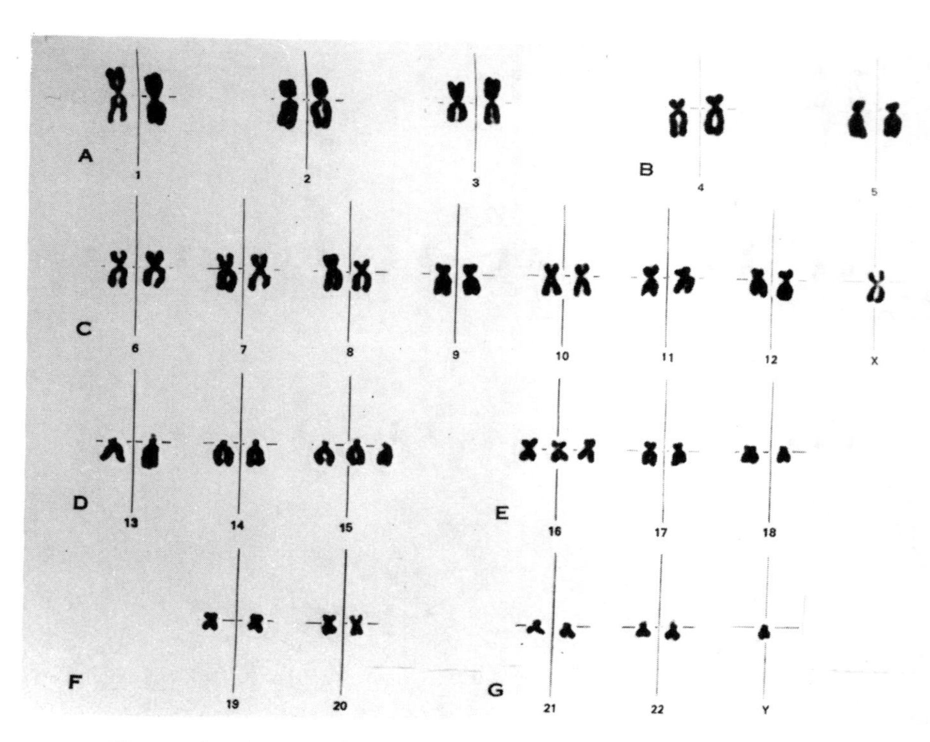

Figure 2: *A case of double trisomy i.e. trisomy 15 and 16.*

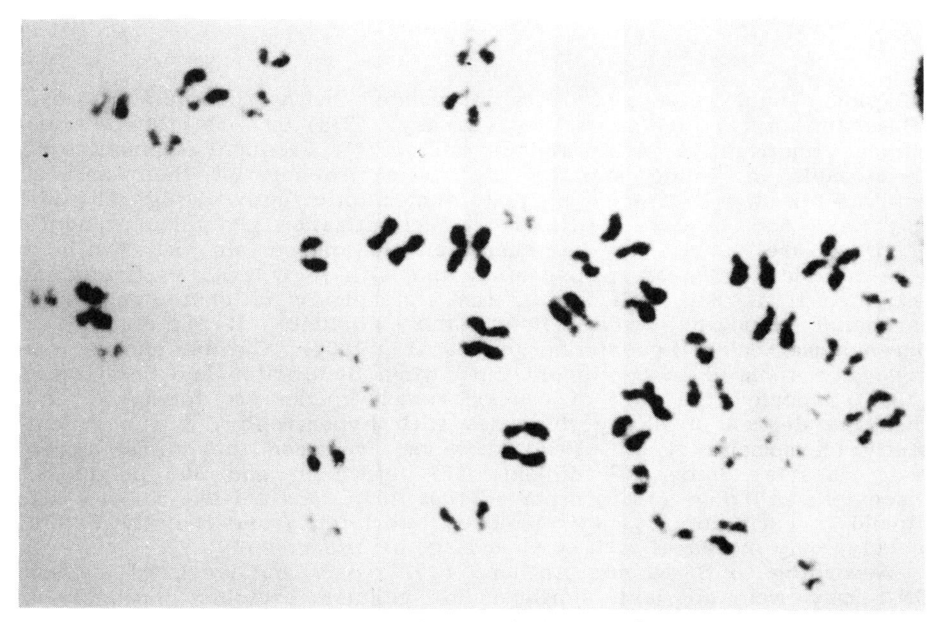

Figure 3: *Chromatid separation.*

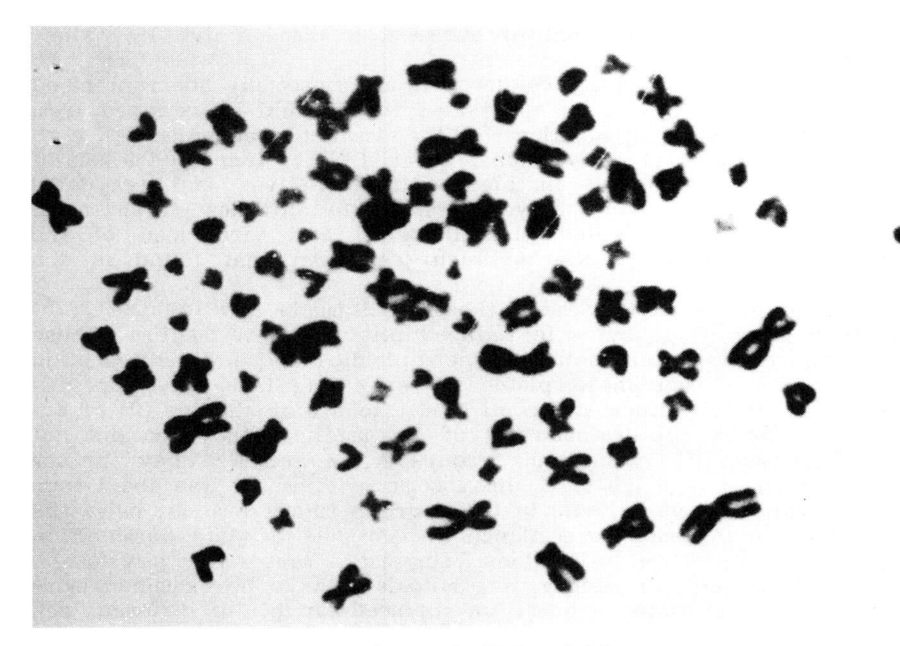

Figure 4: *Tetraploidy.*

DISCUSSION

Polyploidy and aneuploidy are associated with malignant disease (Hamerton, 1971), foetal wastage (Creasy, 1976) or with other serious clinical abnormalities (Atnip and Summit, 1971). Frequent occurrence of these cells in young and healthy native soldiers at their natural habitats, on the contrary, suggests that mitotic faults at high altitude may be a part of normal process of acclimatization. Polyploid amounts of DNA are known to be characteristic of certain cells with a considerable degree of specialisation and with high levels of functional activity. It is now well known that tetraploidy is characteristic of cerebellar Purkinje cells (Bohn and Mitchell, 1976) and liver parenchymal cells (Leuchtenberger et al., 1951). Cardiac muscle was generally assumed to be diploid until when Sandwriter and Scomazzoni (1964) reported that 97% of cardiac muscle nuclei are tetraploid and that the degree of ploidy increases with hypertrophy. A subsequent study (Kompmann et al., 1966), however, revealed the normal human heart muscle to be 20% diploid, 50% tetraploid and 30% polyploid. Eisenstein and Weid (1970) reported that in humans cardiac muscles are diploid in early infancy, tetraploid by late infancy and that the extent of polyploidy increases with advanced age or hypertrophy.

According to Stere and Anthony (1977) polyploid levels of nuclear DNA may well augment synthesis of cellular proteins, probably a necessity in cardiac muscle with rapid turn over of myofibrils and mitochondria and need for large quantities of respiratory enzymes. Polyploidy would also provide a built-in redundancy of DNA available for transcription in the event of chromosome damage due to toxins, hypoxia or metabolites.

In the young and healthy soldiers, if chromosomally aberrant blood cells have no special role in the body, these could be removed from the system quickly. Stimulation of the immune mechanism at high altitude was demonstrated by Chohan et al. (1975). Increased levels of IgG and IgA were found in high altitude natives and sea level residents who had spent 2 years at high altitudes (3,692 m to 5,538 m). It may be noted that bacterial and viral load of the environment at high altitude is much less than that found in the plains.

Our findings closely resemble those of Shimada and Ingalls (1975) when they grew lymphocytes in reduced pH of culture medium. These authors observed hypodiploidies, hyperdiploidies and endoreduplication in nearly 25% of metaphase plates of replicating human lymphocytes exposed in vitro to either acidic pH of 6.5 to 6.9 or alkaline pH of 8.4 to 8.8 equilibrium. In reduced pH of the medium, they too did not observe trisomies in chromosome groups A, B and F. They further report that the frequency of G chromosome involvement was about four times the expected value, and in the E group this was about twice the expectancy. Such close resemblance of two sets of data obtained in different environmental situations suggests that the physiologic environment of the cells undergoing mitosis in vivo be examined more closely at high altitude. Studies on intracellular pH of different cell types of humans are required to justify the existence of intracellular acidosis. There is evidence to show that muscle pH in the Chinese golden hamster declines significantly on hypoxic exposure equivalent to 10,600 m altitude (Shimada et al., 1980).

Vogel and Motulsky (1977) maintain that anaphase lagging and chromosomal non-disjunction are the principal mechanisms which lead to numerical abnormalities, and that acrocentric chromosomes run a higher risk of being involved. Our observations concerning these aspects at high altitude support the above contentions, as most of the observed trisomies and monosomies pertain to the D, E and G groups. After allowing for random chromosome loss, metaphases showing group G monosomies still exceed their trisomic counterparts. This indicates that chromosome lagging is more common in the G group. In the D group, trisomies prevail after allowing for random chromosome loss, indicating a preference for chromosomal non-disjunction.

REFERENCES

ATNIP, R.L. and SUMMIT, R.L. (1971): Tetraploidy and 18 trisomy in a six year old triple mosaic boy. Cytogenetics, 10: 305-317.

BOHN, R.C. and MITCHELL, R.B. (1976): Cytophotometric identification of tetraploid Purkinje cells in young and aged rats. J. Neurobiol., 7: 255-258.

CHOHAN, I.S., SINGH, I., BALAKRISHNAN, K., and TALWAR, G.P. (1975): Immune response in human subjects at high altitude. Int. J. Biometeor., 19: 137-143.

CREASY, M.R., CROLLA, J.A., and ALBERMAN, E.D. (1976): A cytogenetic study of human spontaneous abortions using banding techniques. Human Genet., 31: 177-196.

EISENSTEIN, R., and WEID, G.L. (1970): Myocardial DNA and protein in maturing and hypertrophied hearts. Proc. Soc. Exptl. Biol. Med., 133: 176-179.

GHOSH, P.K. and SINGH, I.P. (1974): Cytogenetic studies in lymphocytes from normal human subjects. The Anthropologist, XIX 1-8.

HAMERTON; J.L. (1971): Human cytogenetics. Academic Press, New York and London, p. 435-441.

HIGURASI, M., and CONEN, P.E. (1971): Comparison of chromosomal behaviour in cultured lymphocytes and fibroblasts from patients with chromosomal disorders and controls. Cytogenetics (Basel), 273-285.

INGALLS, T.H., CURLEY, F.J., and PRINDLE, R.A. (1950): Anoxia as a cause of foetal death and congenital defect in the mouse. Amer. J. Dis. Child, 80: 34-45.

INGALLS, T.H. and YAMAMOTO, M. (1972): Hypoxia as a chromosomal mutagen. Arch. Environ. Health, 22: 305-315.

INGALLS, T.H., SHIMADA, T., and YAMAMOTO, M. (1976): Hypoxia and induced mutations in Syrian Golden hamsters. Arch. Environ. Health, 31: 153-159.

JACOBS, P.A., COURT BROWN, W.M., and DOLL, R. (1964): Distribution of human chromosome counts in relation to age. Nature, 191: 1178.

KNAUTILA, S., SIMELL, O., LIPPONEN, P., and SAARINEN, I. (1976): Bone marrow chromosomes in healthy subjects. Hereditas, 82: 29-36.

KOMPMANN, M., PADDAGS, I., and SANDWRITTER, W. (1966): Feulgen cytophotometric DNA determinations on human hearts. Arch. Pathol., 82: 303-308.

LEUCHTENBERGER, C., VENDRELY, R., and VENDRELY, C. (1951): A comparison of the content of deoxyribose nucleic acid (DNA) in isolated animal nuclei by cytochemical and chemical methods. Proc. Natl. Acad. Sci., 37: 33-38.

LITTLEFIELD, L.G. and GOH, K.O. (1973): Cytogenetic studies in control men and women: Variation in aberration frequencies in 29709 metaphases from 305 cultures obtained over a three year period. Cytogenet. Cell Genet., 12: 17-34.
LUBS, H.A. and RUDDLE, F.M. (1970): Applications of quantitative karyotype to chromosome variation in 4400 consecutive new borns. In: Human population cytogenetics.
MUELLER, H.J. (1971): Variations in karyotypes of normal premature new born babies and infants. Human Genetik, 14: 33-43.
SANDWRITTER, W., and SCOMAZZONI, G. (1964): Deoxyribose nucleic acid content (Feulgen photometry and dry weight) interference microscopy of normal and hypertrophic heart muscle fibres. Nature, 202: 100-101.
SHIMADA, T., and INGALLS, T.H. (1975): Chromosome mutations and pH disturbances. Arch. Environ. Health, 30: 196-200.
SHIMADA, T., WATANBE, G., and INGALLS, T.H. (1980): Trisomies and Triploidies in hamster embryos: Induction by low pressure hypoxia and pH imbalances. Arch. Environ. Health, 35: 101-105.
STERE, A.J., and ANTHONY, A. (1977): Myocardial Feulgen: DNA levels and capillary vascularisation in hypoxia exposed rats. J. Appl. Physiol.: Respirat. Environ. Exercise Physiol., 42: 501-507.
VOGEL, F., and MOTULSKY, A.G. (1977): Human chromosomes. In: Human Genetics. Springer Verlag publishers. Berlin. Heidelberg - New York, pp. 38.

IMMUNE RESPONSES IN MAN AT HIGH ALTITUDE

I.S. Chohan
(Specialist in Pathology & Biochemistry, Mil Hosp
(Cardiothoracic Centre), Pune - 411 040, India)

Abstract: - Both immune responses, humoral and cell mediated immunity (CMI), prevail at a higher plane at high altitude (HA) than at sea level. A subtle evidence exists that HA effect itself is of immunogenic origin and HA exerts a synergistic action with immunogenic stimulus. There is a dichotomy of immune responses in patients of high altitude pulmonary oedema; IgG, IgA and IgM are elevated on one hand and CMI is impaired on the other. CMI remains insufficient, initially, in patients of frostbite and chilblain but is restored fully on recovery. In disorders of chronic mountain sickness, systemic hypertension and asthma, CMI is adequate throughout the stay at HA. While CMI is adequate in patients of chicken pox and mumps, it is depressed in those of infectious hepatitis, hepatic amoebiasis, and malaria in acute stage; however, recovery is fast at HA. Benefits which accrue from a stay at HA, result from acceleration of immune responses and fibrinolytic system. Augmentation of immune responses at HA has, therefore a therapeutic potential.

INTRODUCTION

A great deal of work on physiological and pathological aspects of human biology has been accomplished at high altitudes. Much remains to be done about biological effects of high altitude (HA) on immune responses. HA adapted guinea pigs acquire an increased resistance to virus challenge and show a concomitant rise in beta and gamma globulins (Trapani, 1966; 1969). In human subjects both primary and secondary immune responses are accelerated (Chohan et al., 1975) and cell-mediated immunity (CMI) is equally augmented in temporary residents and natives at HA (Chohan and Singh, 1979). This communication presents a resume of immune responses at HA in human subjects and results of our extended work in various disorders at HA.

MATERIAL AND METHODS

In all 356 subjects were studied, of whom 109 for humoral immune response and 247 for CMI. All were adults between the ages of 17 and

45 years. Immunoglobulins IgG, IgA and IgM were estimated by the Mancini technique modified by Fahey and McKelvey (1965) in 31 healthy sea-level residents; 18 temporary residents, plainsmen inducted to 3,692 m in the western Himalayas and staying there for 2 years; 12 recent arrivals within 7 to 10 days; and 10 permanent natives born and brought up at HA. Of the patients, 19 were suffering from high altitude pulmonary oedema (HAPO) staying at altitudes between 3,692 and 5,538 m; 15 were temporary residents who sustained frostbite; and 4 were high-altitude natives who were carriers of Australia antigen (HBsAg).

To ascertain the status of CMI, the parameters included were: haemoglobin, haematocrit, total and differential leucocyte counts, and absolute lymphocyte counts determined by standard haematological procedures; T-rosettes (Jondal et al., 1972); B-rosettes (Stjernsward et al., 1972); PHA phytohaemagglutinin-blast transformation of lymphocytes (Pentycross, 1968); Lymphocyte migration index (LMI) (Federlin et al., 1971); and DNCB (1-chloro-2, 4-dinitrobenzene)-response (Catalona et al., 1972).

DNCB-response was measured in all 247 subjects. These included 42 sea-level residents, 42 temporary residents, and 45 high altitude-natives (24 males, 21 females). Of the patients, 12 were of HAPO, 20 of chilblain, 15 of frostbite,14 of systemic hypertension, 6 each of chronic mountain sickness (CMS), asthma, chicken pox and malaria, 5 of mumps, 17 of infectious hepatitis, 10 of hepatic amoebiasis, and a single case of malignancy (primary hepatoma) in a high altitude native female of 65 years of age. Skin biopsies from the site of DNCB-application were studied for histological evidence of intensity of DNCB reaction in normal subjects and patients.

Parameters of CMI, other than DNCB-response, were carried out in 24 sea-level residents, 24 temporary residents, 24 natives of HA and in patients including 10 of HAPO, 5 of infectious hepatitis, 5 of hepatic amoebiasis, and 4 of malaria. However, only haemoglobin, haematocrit, total and differential leucocyte counts, besides DNCB-response, were carried out in patients of frostbite (6), systemic hypertension (14), CMS (6), asthma (6), chicken pox (6), mumps (5) and malignancy (1).

Apart from the above investigations, cellular magnesium levels were determined by using Perkin-Elmer atomic absorption spectrophotometer in the RBCs and WBCs of 23 sea level residents, 22 high altitude-natives and 6 patients of high altitude pulmonary oedema.

RESULTS

Results are summarised in Table 1, 2, and 3 and Fig. 1 and 2.

Immunoglobulins (Table 1): IgG, IgA and IgM were significantly higher in recent arrivals at high altitude compared with sea-level residents. IgG and IgA were significantly higher in temporary residents and natives at HA, but IgM though numerically higher was not significantly increased. IgG, IgA, and IgM showed a significantly marked increase in

TABLE 1: *Immunoglobulins levels at high altitude.*

Subjects	Serum immunoglobulins					
	IgG mg/dl		IgA mg/dl		IgM mg/dl	
	Mean	SD	Mean	SD	Mean	SD
Sea level residents (31)	1140	254	205	50	156	50
Recent arrivals (12)	4285	140**	369	74**	207	65*
High altitude pulmonary oedema patients (19)	6975	1015**	755	240**	690	247**
Temporary residents at high altitude (18)	1777	93**	292	23**	185	27
Natives at high altitude (HBsAg negative) (10)	1742	515**	288	37**	184	65
Natives at high altitude (HBsAg positive) (4)	2404	363**	393	140**	196	107
Frostbite patients (15)	5197	532**	1136	226**	125	20*

Parentheses indicate the number of subjects tested.
* $P < 0.05$, ** $P < 0.01$.

patients of high altitude pulmonary oedema compared with sea-level residents, temporary residents, natives, and recent arrivals at high altitude. Natives who were carriers of Australia antigen (HBsAg positive) had significantly higher levels of all three immunoglobulins compared with temporary residents and healthy natives of HA. Frostbite subjects had significantly increased IgG and IgA, but IgM showed a significant decrease in them compared with sea level residents, temporary residents at HA and healthy natives.

DNCB-Response (Table 2, Fig. 1 and 2): DNCB-response was higher in temporary residents and natives of whom 100% and 95.6% became positive respectively compared to 78.6% sea-level residents. "Spontaneous flare" (4+,3+) reaction also in temporary residents and natives was about 93% and 87% respectively compared to 62% in sea-level residents. Only 4.4% natives remained non-responders compared to 21.4% sea-level residents.

DNCB-response was conspicuous by its absence in the acute stage of illness in HAPO, however, it reappeared as 'Spontaneous flare' in 4 out of 6 patients within 2 weeks. All subjects of frostbite and 18 of 20 subjects of chilblain showed 'Spontaneous flare' though the majority had 3+ score. Patients of CMS, systemic hypertension and asthma exhibited a high score of 4+ with one exception each in the latter categories. Among the infection group of patients those of chicken pox and mumps showed a 'Spontaneous flare', however, patients of

TABLE 2: *Summary of DNCB - Response at high altitude.*

	CMI to DNCB - Reaction				
	Positive				Negative
Subjects	4+	3+	2+	1+	0
Sea level residents (42)	7	19	5	2	9
Temporary residents at HA(42)	31	8	2	1	-
Natives at HA (45)	27	12	2	2	2
High altitude pulmonary oedema: -					
Acute (6)	-	-	-	-	6
Recovered (6)	1	3	1	1	-
Chilblain, Acute (20)	7	11	-	2	-
Frostbite, Acute (15)	1	14	-	-	-
Systemic hypertension (14)	13	-	-	1	-
Chronic mountain sickness (6)	6	-	-	-	-
Asthma (6)	4	1	-	1	-
Chicken pox, Acute (6)	6	-	-	-	-
Mumps, Acute (5)	5	-	-	-	-
Infectious hepatitis, Acute (5)	-	-	2	3	-
- do - , Recovered (12)	3	9	-	-	-
Hepatic amoebiasis, Acute (5)	-	-	3	2	-
- do - , Recovered (5)	3	2	-	-	-
Malaria, Acute (2)	-	-	1	1	-
-doo, Recovered (4)	2	2	-	-	-
Primary hepatoma (1)	-	-	-	-	1

Parentheses indicate the number of subjects. HA = High altitude.

infectious hepatitis, hepatic amoebiasis and malaria, in acute stage of illness, who registered a low score of 2+ and 1+(on challenge), made a fast recovery within 2 weeks with a high score of 'Spontaneous flare'. The case of malignancy did not respond despite a challenge and remained negative even on rechallenge.

Skin biopsies from the site of DNCB application showed a reaction of higher intensity at HA in both temporary residents and natives. Intense erythema, acanthosis and multiple bullae formation in epidermis and lymphocytic infiltration in the perivascular regions of dermis were prominent features in them (Fig. 1). In contrast, skin reaction in the acute stage of HAPO showed scant or next to nil lymphocytic infiltration in dermis (Fig. 2), however, it appeared slowly on

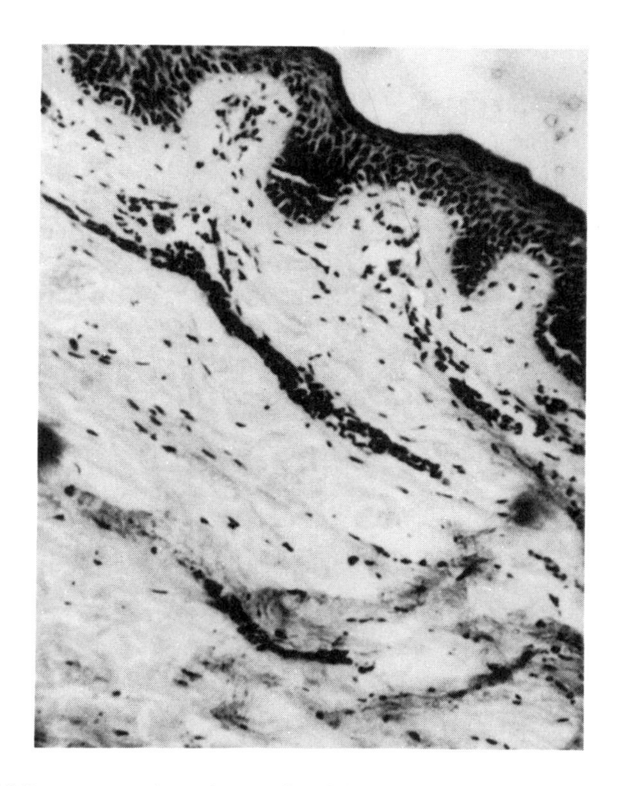

Figure 1: *DNCB - reaction in a healthy individual at high altitude depicting acanthosis in epidermis and widespread and profuse lymphocytic infiltration in the dermis.*

recovery. Lymphocytic infiltration of dermis was adequate in other disorders including CMS, systemic hypertension, asthma, frostbite, chilblain (commensurate with 3+ DNCB-response), chicken pox and mumps. In infectious hepatitis, hepatic amoebiasis and malaria, the lymphocytic response was consistent with process of recovery.

Haemoglobin (Table 3): Compared to sea-level residents haemoglobin increased significantly in temporary residents but showed no difference in high altitude natives. In HAPO and asthma patients haemoglobin showed a significant decrease compared to temporary residents though it was still higher than in sea-level residents and natives at HA. In patients of CMS and systemic hypertension haemoglobin showed a significantly marked rise compared to all the categories at HA. In hepatic amoebiasis it was at par with temporary residents and was the lowest recorded in the malignant hepatoma case. In other patients of frostbite, chicken pox, mumps, infectious hepatitis and malaria, haemoglobin was significantly lower compared to temporary residents

Figure 2: *DNCB - reaction in a patient of high altitude pulmonary oedema during acute illness showing scanty lymphocytic response in the dermis.*

though it did not show significant difference compared to sea-level residents and natives.

Haematocrit (Table 3): Compared to sea-level residents haematocrit was significantly raised both in temporary residents and natives at HA and it was still higher in the former. Haematocrit decreased significantly in patients of HAPO, though the difference was not significant compared to sea-level residents and natives. CMS patients had the highest haematocrit (74.3 ± 3.1) compared to all the categories at HA whereas the malignant case had the lowest haematocrit values (38.5 ± 1.6). In patients of frostbite, systemic hypertension and asthma, haematocrit decreased significantly compared to the temporary residents, though the values were significantly higher compared to sea-level residents and natives. Haematocrit decreased significantly in patients of chicken pox, mumps, infectious hepatitis, hepatic amoebiasis, and malaria compared to temporary residents, though it was at par with that of natives and only mildly raised compared to sea-level residents.

TABLE 3: *Parameters of Cell mediated immune-Response at high altitude.*

	Haemoglobin (g/dl)		Haematocrit (%)		Total Leucocyte count/cmm		Absolute Lymphocyte count/cmm		T-Rosettes %		B-Rosettes %		PHA-Blasts %		Lymphocyte migration index (LMI)	
	Mean	SD	Mean	SD	Mean	SD	Mean	SD	Mean	SD	Mean	SD	Mean	SD	Mean	SD
Sea-level residents (24)	14.1	0.16	43.5	0.8	7600	430	2300	108	53	2	14.5	4	53	4	1.0	0.2
Temporary residents at HA (24)	16.2	0.22**	53.1	0.8**	7200	402**	2290	99	46	4	22.0	4***	66	4***	4.4	0.4**
Natives at HA (24)	13.6	0.14	47.0	0.9*	6275	204***	2015	84*	43	5	22.0	2***	64	2***	4.1	0.4**
High altitude pulmonary oedema(10)	15.6	1.40*	43.6	5.3***	10190	402***	2200	85	10	4***	23.0	5***	49	8**	0.6	0.1***
Frostbite (6)	13.9	1.20*	48.5	0.8***	9150	409***	-		-		-		-		-	
Systemic hypertension (14)	16.7	0.63*	50.2	1.8***	9075	495***	-		-		-		-		-	
Chronic mountain sickness (6)	22.6	0.99***	74.3	3.1***	8600	470***	-		-		-		-		-	
Asthma (6)	15.6	0.26*	48.0	0.6***	8675	220***	-		-		-		-		-	
Chicken pox (6)	14.4	1.40***	45.8	1.9***	8900	415***	-		-		-		-		-	
Mumps (5)	14.3	1.31***	45.4	1.8***	8679	419***	-		-		-		-		-	
Infectious hepatitis (5)	14.5	1.48***	46.5	2.1***	8800	430***	2412	98	16	4**	26.0	2***	54	4	4.7	1.9**
Hepatic amoebiasis (5)	15.4	0.80	48.3	2.1***	6400	110***	2109	58*	13	2***	21.0	4**	56	9	1.8	0.6***
Malaria (4)	13.9	1.54***	45.6	2.0***	6100	495***	1821	85***	10	3***	17.0	2*	54	4	2.3	0.8**
@Malignancy-primary hepatoma(1)	8.5	0.65	38.5	1.6	6200	190	-		-		-		-		-	

Parentheses show the number of subjects/patients studied.

* $P < 0.01$, ** $P < 0.005$, *** $P < 0.001$.

@ Average of four readings have been taken to calculate Mean and SD in this case.

Note: Parameters of Temporary residents and Natives at high altitude have been compared with those of Sea level residents, and those of patients have been compared with the Temporary residents (high-altitude controls) and Sea level residents (sea-level controls).

Total leucocyte and absolute lymphocyte count (Table 3): Compared to sea-level residents total leucocyte counts were significantly lowered in temporary residents ($P < 0.005$) and natives ($P < 0.001$) but absolute lymphocyte counts decreased significantly only in natives ($P < 0.01$). The differential leucocyte count was similar in both and was accompanied with a mild fall in monocytes and a significant rise in eosinophils. In patients of HAPO a significant rise occurred in total leucocyte count but they had a significant fall in absolute lymphocyte count ($P < 0.001$) compared to sea-level residents. Patients of frostbite, systemic hypertension, CMS and asthma also had a significant rise in total leucocyte count but in differential count they had significant fall in lymphocytes and relative rise in polymorphs, monocytes and eosinophils compared to temporary residents. Compared to temporary residents and natives, total leucocyte count showed a significant increase in patients of chicken pox, mumps, and infectious hepatitis but a significant decrease in patients of hepatic amoebiasis, malaria, and malignancy; absolute lymphocyte count showed a significant fall in hepatic amoebiasis and malaria; however, differentially monocytes and eosinophils showed a rise in patients of infectious hepatitis, hepatic amoebiasis and malaria.

T and B-Rosettes (Table 3): Compared to sea-level residents T-rosettes did not show any significant change in temporary residents and natives at HA but there was a significant increase in B-rosettes ($P < 0.001$) in both of them. Patients of HAPO showed a significant decrease in T-rosettes and a significant increase in B-rosettes ($P < 0.001$) during the acute stage of illness compared to sea-level residents. Similarly patients of infectious hepatitis, hepatic amoebiasis, and malaria showed a significant decrease in T-rosettes and increase in B-rosettes in them in the acute stage of illness. The rise in number of B-rosettes was the highest recorded among the patients of infectious hepatitis.

PHA-blast transformation of lymphocytes (Table 3): Compared to sea-level residents PHA-blasts of lymphocytes increased significantly in both temporary residents and natives at HA ($P < 0.001$). In patients of HAPO, PHA-blasts were significantly declined ($P < 0.005$) in the acute stage of the illness. Compared to sea-level residents PHA-blasts in patients of infectious hepatitis, hepatic amoebiasis, and malaria were numerically lower but not significantly different compared to temporary resident and natives. PHA-blasts showed a mild but significant decrease in patients of infectious hepatitis and malaria ($P < 0.05$).

Lymphocyte migration index (Table 3): Compared to sea-level residents the increase in LMI was highly significant in both temporary residents and natives at HA. LMI was significantly low ($P < 0.001$) in patients of HAPO compared to temporary residents and natives at HA, the change, however, was insignificant compared to the sea-level residents in acute stage of illness, but on recovery, LMI was regained fast in all patients and became comparable to that of natives and temporary residents. Patients of infectious hepatitis did not show any decline in LMI which remained as high as in temporary residents and natives. LMI decreased significantly in hepatic amoebiasis and malaria patients compared to temporary residents and natives, in acute stages.

Magnesium levels: Mg^{++} contents of RBCs and WBCs expressed in mM/L (Mean ± SD) are tabulated below:

	RBC	WBC	WBC/RBC
Sea-level residents (23)	1.670 ± 0.117	1.955 ± 0.272	1.170
	Vs		
Natives at high altitude(22)	1.913 ± 0.142***	2.007 ± 0.126***	1.049
	Vs		
High altitude pulmonary oedema patients (6)	1.795 ± 0.127**	1.786 ± 0.101**	0.994

Compared to sea-level residents both erythrocytic and leucocytic magnesium contents were significantly increased in healthy natives resident at high altitude ($P < 0.001$). Both erythrocytic and leucocytic magnesium contents were significantly lower in patients of high-altitude pulmonary oedema ($P < 0.005$) compared to natives, though these were not significantly different compared to sea-level residents. However, the ratio of leucocyte to erythrocyte magnesium level remained more than unity in sea-level residents and natives at HA, the values being 1.170 and 1.049 respectively, this ratio was reduced to 0.994 in patients of HAPO.

DISCUSSION

There is a quantitative change in immunoglobulin synthesis on exposure to HA. On arrival at HA when the stress is greatest the immunoglobulin synthesis is also at its peak. This is evident in the recent arrivals in whom IgG, IgA, and IgM are markedly higher in the absence of any antigenic stimulus. However, synthesis of immunoglobulins is augmented if there is an antigenic challenge (Chohan et al., 1975) or occurrence of high altitude pulmonary oedema, HBsAg carrier state, or frostbite. On adaptation, levels of these immunoglobulins decline to a large extent but still remain higher as seen in temporary residents and healthy natives at HA. These results indicate that HA itself exerts an immunogenic stimulus and provides a synergistic effect in the presence of immunogenic challenge. These observations confirm the belief of Tengerdy and Kramer (1968) on augmentation of humoral immune response at or under simulated HA conditions.

CMI, the other arm of immune response, is equally accelerated at HA, despite a relative fall of total leucocyte count in the temporary residents and natives and of absolute lymphocyte count in the natives. The qualitatively augmented functions of lymphocytes at HA are reflected by an increase in T-rosettes, PHA-blasts transformation of lymphocytes, LMI and DNCB- response, and are also histologically evident. A rise in B cell population at HA is obviously responsible for higher synthesis of immunoglobulins. From the above account it is clear that both CMI and humoral immune responses prevail at higher plane and immune-surveillance is more potent at HA than at sea level. Possible mechanism of augmentation of immune responses at HA has been discussed by us elsewhere (Chohan et al., 1975, Chohan and Singh, 1979), though it remains unclear how HA influences this system. Higher haematocrit appears more relevant in influencing CMI favourably (Chohan and Singh, 1979).

An interesting finding of this study is the significantly higher erythrocyte and leucocyte magnesium levels in residents at HA. The role of magnesium influencing CMI and humoral immunity is well documented. Magnesium deficiency is associated with depressed synthesis of immunoglobulins (Alcock and Shils, 1974; Mccoy and Kenney, 1975) and depressed CMI as a result of impaired thymus growth (Bois, Sandborn and Messier, 1969). Higher magnesium levels in plasma are also related to higher haematocrit (Whang and Wagner, 1966). Therefore, increased magnesium contents in erythrocytes and specially in leucocytes have a direct relevance of influencing immune responses favourably at HA. Another interesting effect which magnesium exerts is on blood coagulation which is reduced by its vasodilatory effect, its stabilisation of fibrinogen and platelets, and by promotion of fibrinolysis (Szelenyi, 1971; Szelenyi et al., 1967). On arrival at HA fibrinolytic activity is normally increased and remains high during the stay at HA (Chohan, Singh and Balakrishnan, 1974). Magnesium may, therefore, contribute to augment immune responses and enhance fibrinolytic activity at HA.

High altitude provides both enhanced fibrinolytic activity and accelerated immune responses and hence offers an effective immune clearance. Most of the benefits which accrue from staying at HA could be attributed to this potential, and rarity of malignancy at HA can, perhaps, be explained on this basis (Singh et al., 1977).

The following paragraphs briefly highlight the role of immune response in various disorders at HA. In HA pulmonary oedema diminished plasma fibrinolytic activity and increased intravascular coagulation are the hall mark in the acute stage of illness (Singh and Chohan, 1974). Immune deviation in this disorder was pointed out when reduced electrophoretic mobility of platelets was found associated with increased levels of IgG and IgM (Chohan, 1971; Chohan, Balakrishnan and Talwar, 1971). Dichotomy of immune apparatus in this disorder becomes more evident now as immunoglobulins are raised on one hand and CMI is impaired on the other hand. Reduced magnesium levels in erythrocytes and leucocytes of patients of HAPO are further corroborative evidence of impaired immune responses in this illness. At present in our experience oral administration of 20 mg of furosemide is capable of reversing an adverse coagulation trend by enhancing fibrinolytic activity, inhibiting platelet aggregation, and by accelerating lymphocyte migration without any ill effects (Singh and Chohan, 1973; Chohan et al., 1977; Chohan, Singh and Vermylen, 1977; Chohan, 1980).

In initial stages of frostbite and chilblain, CMI is of lower order as revealed by DNCB-response, however, immunoglobulins IgG and IgA are raised and IgM is reduced in frostbite cases. Intravascular coagulation in these cold injuries is associated with diminished fibrinolytic activity and presence of fibrinogen degradation products and cryoglobulins (Chohan, Singh and Balakrishnan, 1975; 1978). A larger percentage of cryoglobulins belongs to IgM which is consumed in the process of coagulation. Inhibition of fibrinolytic activity is known to block the reticuloendothelial system (Lee, 1962), therefore, in the presence of insufficient CMI, persistence of FDPs and cryoglobulins is likely to aggravate intravascular coagulation in early stages of these disorders.

Patients of systemic hypertension, chronic mountain sickness, and asthma at HA possess an adequate degree of CMI as shown by the 'Spontaneous flare' phenomenon of DNCB-response in these cases. Total leucocyte count, haemoglobin and haematocrit in them remain high commensurate with HA environments, and lymphocytic infiltration and their recruitment in the dermis after DNCB application is plenty as seen in histological sections. CMS in the western Himalayas has been discovered recently (Nath et al., 1983). Among all the patients at HA, CMS is associated with highest haemoglobin and the haematocrit and all the plethora of symptoms and signs of chronic alveolar hypoventilation. However, one redeeming feature in this disorder is that it is associated with increased plasma fibrinolytic activity (our unpublished data) which takes care of the blood viscosity. Perhaps, this is the reason for its protracted course to develop. Low incidence of systemic hypertension and 'cure' of asthma at HA (Singh et al., 1977; Tromp and Bouma, 1974) may partly be due to adequate CMI apart from other factors such as stimulation of adrenal cortex and fibrinolytic activity at HA.

In the infection group CMI potential is higher in patients of chicken pox and mumps, whereas it is of low order in patients of infectious hepatitis, hepatic amoebiasis and malaria in acute stage of illness. However, patients of the latter categories make a fast recovery at HA. Low incidence of above diseases at HA is understandable not only due to hindrance of active transmission under low temperature but also because of effective handling of the offending organism by potent mechanisms of fibrinolysis and immune responses at HA (Singh et al., 1977). Intravascular coagulation which may invariably occur during viral, bacterial and protozoal infections protects these organisms in various organs and is responsible for severity, resistance and relapse. Countering coagulation, fibrinolytic system may afford uncovering and elimination of the infectious organisms by increased immune response at HA. This perhaps is the cause for relapses of malaria and flare up of hepatic amoebiasis at HA (Singh et al., 1977). The augmentation of immune apparatus and fibrinolytic system at high altitude has therapeutic potential.

REFERENCES

ALCOCK, N.W. and SHILS, M.E. (1974): Serum immunoglobulin G in the magnesium depleted rats. Proc. Soc. exp. Biol. (N.Y.), 145: 855-858.

BOIS, P., SANDBORN, E.B., and MESSIER, P.E. (1969): A study of thymic lymphosarcoma developing in magnesium deficient rats. Cancer Res., 29: 763-775.

CATALONA, W.J., TAYLOR, P.T., ROBSON, A.S., and CHRETIEN, P.B. (1972): A method for DNCB contact sensitization: A clinicopathological study. New Eng. J. Med., 286: 399-402.

CHOHAN, I.S. (1971): Fibrinolytic activity, platelet functions and immune response at high altitude. Thesis for MD, All India Institute of Medical Sciences, New Delhi.

CHOHAN, I.S. (1980): First few days at high altitude and furosemide. Int. Rev. Army, Navy and Air Force Med. Services, 53: 745-753.

CHOHAN, I.S. and SINGH, I. (1979): Cell mediated immunity at high altitude. Int. J. Biometeor., 23: 21-30.

CHOHAN, I.S., BALAKRISHNAN, K. and TALWAR, G.P. (1971): Electrophoretic mobility of platelets in high altitude pulmonary oedema and high altitude pulmonary hypertension, and immunoglobulins levels at high altitudes. Armed Forces Med. J. (India), 27: 437-450.

CHOHAN, I.S., SINGH, I., and BALAKRISHNAN, K. (1974): Fibrinolytic activity at high altitude and sodium acetate buffer. Throm. Diath. haemorrh., 32: 65-70.

CHOHAN, I.S., SINGH, I., and BALAKRISHNAN, K. (1975): Blood coagulation and immunological changes associated with frostbite at high altitude. Int. J. Biometeor., 19: 144-154.

CHOHAN, I.S., SINGH, I., and BALAKRISHNAN, K. (1978): Frostbite at high altitude: Immunological, biochemical and blood coagulation changes during its occurrence. Aspects Allergy Appl. Immunol., X: 135-148.

CHOHAN, I.S., SINGH, I., BALAKRISHNAN, K., and TALWAR, G.P. (1975): Immune response in human subjects at high altitude. Int. J. Biometeor., 19: 137-143.

CHOHAN, I.S., SINGH, I., and VERMYLEN, J. (1977): Furosemide: Its role on certain platelet functions. Acta Clinica Belgica, 32: 39-43.

CHOHAN, I.S., SINGH, I., VERMYLEN, J., and VERSTRAETE, M. (1977): Effect of furosemide on plasma fibrinolytic activity and urokinase excretion. Exp. Hematol., 5: 153-157.

FAHEY, J.L., and McKELVEY, E.M. (1965): Quantitative determination of serum immunoglobulins in antibody agar plates. J. Immunol., 94: 84-90.

FEDERLIN, K., MAINI, R.N., RUSSEL, A.S., and DUMONDE, D.C. (1971): A micromethod for peripheral leucocyte migration in tuberculin-sensitivity. J. clin. Path., 24: 553-556.

JONDAL, M., HOLM, G., and WIGZELL, H. (1972): Surface markers on human T and B lymphocytes. 1. A large population of lymphocytes forming non-immune rosettes with sheep red cells. J. Exp. Med., 136: 207-215.

LEE, L. (1962): Reticuloendothelial clearance of circulating fibrin in the pathogenesis of the generalized Schwartzman reaction. J. Exp. Med., 115: 1065-1082.

MCCOY, J.H., and KENNEY, M.A. (1975): Depressed immune response in the magnesium-deficient rats. J. Nutr., 105: 791-797.

NATH, C.S., KASHYAP, S.S., and SUBRAMANIAN, A.R. (1983): Chronic mountain sickness - A therapeutic trial at high altitude. Armed Forces Med. J. (India), 39: 131-137.

PENTYCROSS, C.R. (1968): Technique for lymphocyte transformation. J. clin. Path., 21: 175-178.

SINGH, I., and CHOHAN, I.S. (1973): Reversal of abnormal fibrinolytic activity, blood coagulation factors and platelet function in high pulmonary oedema with frusemide. Int. J. Biometeor., 17: 73-81.

SINGH, I. and CHOHAN, I.S. (1974): Adverse changes in fibrinolysis, blood coagulation and platelet function in high altitude pulmonary oedema and their role in its pathogenesis. Int. J. Biometeor., 18: 33-45.

SINGH, I., CHOHAN, I.S., LAL, M., KHANNA, P.K., SRIVASTAVA, M.C., NANDA, R.B., LAMBA, J.S. and MALHOTRA, M.S. (1977): Effects of high altitude stay on the incidence of common diseases in man. Int. J. Biometeor., 21: 93-122.

STJERNSWARD, J., JONDAL, M., VINKY, F., WIGZELL, H., and SEAL, R. (1972): Lymphopenia and change in distribution of human B and T lymphocytes in peripheral blood induced by irradiation for mammary carcinoma. Lancet, II: 1352-1356.

SZELENYI, I. (1971): Physiological interrelationship between magnesium and heart. First Intern. Symp. Magnesium Deficit in Human Pathology, Vittel, France, p. 95.
SZELENYI, I., RIGO, R., AHMEN, B.O., and SOS, J. (1967): The role of magnesium in blood coagulation. Thromb. Diath. Haemorrh. (Stuttg.), 18: 626-631.
TENGERDY, R.P., and KRAMER, T (1968): Immune response of rabbits during short term exposure to high altitude. Nature (Lond.), 217: 367-369.
TRAPANI, I.L. (1966): Altitude, temperature and the immune response. Fed. Proc., 25: 1254-1259.
TRAPANI, I.L. (1969): Environment, infection and immunoglobulin synthesis. Fed. Proc., 28: 1104-1106.
TROMP, S.W., and BOUMA, J.J. (1974): The Biological Effects of Natural and Simulated High Altitude Climate on Physiological functions of Healthy and Diseased subjects (in particular Asthmatics). Monograph Series Biometeor. Res. Centre, XIII, 5-16.
WHANG, R., and WAGNER, R. (1966): The influence of venous occlusion and exercise on serum magnesium concentration. Metabolism, 15: 608-612.

ALTERATIONS IN THYROID GLAND FUNCTION DURING COLD EXPOSURE IN MAN

R.C. Sawhney, A.S. Malhotra, Lazar Mathew
and R.M. Rai
(Defence Institute of Physiology and Allied Sciences,
Delhi Cantt - 110010)

Abstract: - Studies were carried out to investigate changes in thyroid gland function due to acute cold exposure of 10° C, 4 h daily for 6 days in man. Circulatory levels of T_4, T_3, RBC concentration of T_3 and TSH were estimated in eight healthy euthyroid males before and on day 1 and 6 of the cold exposure schedule. Physiological parameters such as heart rate, oral and mean skin temperature, oxygen consumption and shivering activity were studied. Results showed significant increase in plasma T_4 and T_3 on day 1 of exposure and a further rise in T_3 levels on day 6. Thyrotropin levels remained unaltered on day 1 but showed a significant decrease on day 6. T_3 concentration of erythrocytes also showed a decrease immediately after cold exposure but rose to normal levels on day 6. The physiological parameters did not show any significant difference between day 1 and 6 of the exposure. The results of the study showed that cold exposure of moderate severity resulted in an increase in the circulatory levels of thyroid hormones, which was not related to the physiological changes of acclimatization.

INTRODUCTION

Thyroid hormones have their greatest calorigenic action in homeothermic animals and are important in the mechanisms of thermal homeokinesis. The observation that severely hypothyroid animals were unable to survive when the environmental temperature was reduced from 26° C to 16° C (Leblond and Eartly, 1952), suggests that the thyroid is essential for temperature regulation. It is well documented that animals exposed to cold elicit an increased thyroidal uptake of iodine, increased turnover of thyroid hormones (Cottle and Carlson, 1956) and about twofold increase in thyroxine requirement to prevent thyroid hypertrophy (Woods and Carlson, 1956). However, the role of thyroid in adaptation to low temperature in man remains controversial. Increased thyroid hormone turnover (Ingbar and Bass, 1967) and elevated TSH levels (Raud and Odell, 1969) have been demonstrated in man during prolonged cold exposure in the Arctic. However, these observations were not corraborated by other investigators who failed to demonstrate any change in protein-bound iodine (Wilson et al., 1970).

In addition, though a marked rise in extracellular T_4 and T_3 has been recorded in man exposed to cold (Eastman et al., 1974), no attempt has been made to elicit changes in intracellular thyronines. The cytoplasm of the mature erythrocytes in man is a readily sampled intracellular fluid, free of organelles and can be used confidently to monitor intracellular thyroid hormones (Yoshida and Davis, 1980). The present study deals with the initial response of the human thyroid to acute cold exposure and its relationship with physiological responses of short-term acclimatization.

MATERIALS AND METHODS

The study was performed on eight healthy euthyroid male volunteers of the age group 20-26 years, free from renal, hepatic and endocrine dysfunction. None of the subjects studied received medications viz. salicylates, penicillin or dilantin which are known to displace thyroid hormones from binding proteins. They had no previous experience of deliberate exposure to either extreme heat or cold. The cold exposure experiments were performed in September when maximum temperature ranged between 32-39° C and minimum temperature 23-28° C. An elaborate briefing about the test was done to the subjects before initiating the study and their consent was obtained.

Cold exposure was given by exposing the subjects to 10° C for 4 h daily in a thermostatically controlled cold chamber for six days. During the cold exposure, the subjects wore nylon underwears only. The physiological parameters such as heart rate, oxygen consumption, oral temperature and mean skin temperature (computed from four skin sites) were recorded before the cold exposure and after 30, 60 and 120 min of cold exposure on the first day and the sixth day of the schedule. The visible shivering response was also noted and graded during this period.

Heparinized venous blood samples were obtained before, on day 1 and day 6 of cold exposure. Samples were immediately centrifuged for 10 min at 2000 rpm to obtain the cellular pellet and plasma. Plasma was stored at -20° C for the estimation of T_3, T_4 and TSH by radio-immunoassays (Rastogi and Sawhney, 1976; Odell and Fisher, 1971; Rastogi et al., 1973). All the samples from the study were assayed in the same batch to avoid larger inter-assay variations.

T_3 bound to erythrocytes was estimated in the cellular pellet using the technique of Yoshida and Davis (1980). In brief, the pellet was washed thrice with normal saline and lysed with the addition of an equal volume of distilled water and 400 µl of toluene. The lysed pellet was extracted with 5 ml of toluene twice, dried under a stream of nitrogen and reconstituted in assay buffer for T_3 estimation. The standard tubes received an equal amount of hormone-free RBC cytosol (prepared by treating with activated charcoal).

The statistical analyses were perfomed using analysis of variance technique.

RESULTS

Fig. 1 shows circulatory levels of T_4 and T_3, before and during cold exposure. T_3 levels before exposure varied between 109 to 170 ng/dl with a mean of 138.3±8.4 (SE) ng/dl. Plasma T_3 significantly increased ($p < 0.01$) to 264.2±36.5 ng/dl on first day of exposure and showed a further significant increase ($p < 0.05$) to 337±37 ng/dl over the preceding levels on sixth day of cold exposure. Like T_3, T_4 levels also showed a significant rise ($p < 0.01$) from the pre-exposure value of 6.93±0.42 µg/dl to 18.5±1.63 µg/dl on day 1 of cold exposure and remained at higher levels till day 6 (19.0±1.9 µg/dl) of exposure.

Plasma TSH before cold exposure varied between 1.5 to 3.8 µU/ml with a mean of 2.85±0.26 µU/ml. Although TSH levels did not show any significant change ($p > 0.05$) on day 1 of exposure (3.09±0.4), it was

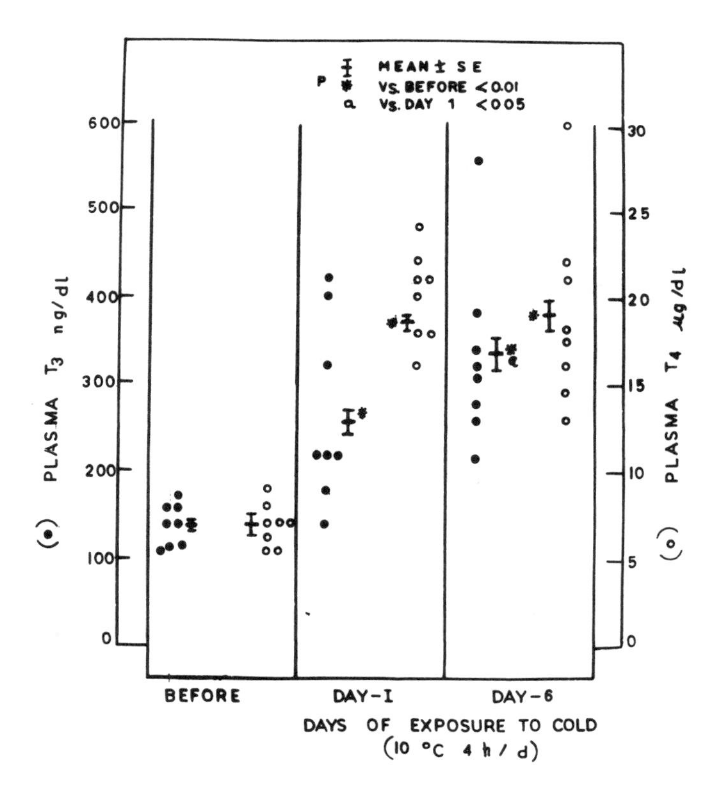

Figure 1: *Plasma levels of T_4 and T_3 before and on day 1 and day 6 of cold exposure.*

significantly decreased ($p < 0.01$) to 1.15±0.36 μU/ml on day 6 (Fig. 2). T_3 concentration of erythrocytes showed a significant decrease ($p < 0.01$) from the pre-exposure value of 18.37±1.87 ng/dl to 10.25±1.46 ng/dl on day 1 of exposure. However, on day 6 of cold exposure, T_3 concentration of erythrocytes returned to pre-exposure levels (16.87±2.22 ng/dl).

The changes in physiological parameters during the 2 h period of cold test on day 1 and day 6 are given in Table 1. There was no significant difference ($p > 0.05$) in the changes observed in any of the parameters between day 1 and day 6. However, the acute cold exposure has resulted in significant ($p < 0.01$) reduction of heart rate, oral and mean skin temperatures by about 60 min of exposure itself. The oxygen consumption showed a steady rise from 30 min of exposure onwards and the shivering became very severe and continuous by the end of the 2 h period of the cold test.

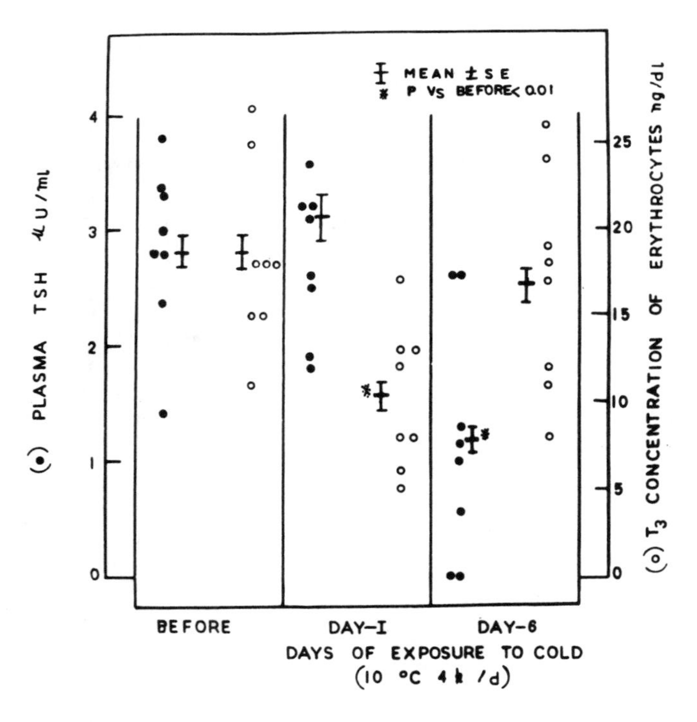

Figure 2: *Alterations in plasma TSH and T_3 concentration of erythrocytes before and during cold exposure.*

TABLE 1: *Mean ± SE of Physiological Parameters before and on day 1 (D1) and six (D6) of acute cold exposure (10° C, 4 h/d).*

Parameters	Before		Time after cold exposure					
			30 min		60 min		120 min	
	D_1	D_6	D_1	D_6	D_1	D_6	D_1	D_6
Heart rate (beats/min)	69.75± 0.59	68.75± 0.64	69.25± 0.45	68.62± 0.49	63.62±* 0.59	63.75* 0.65	62.37±* 0.42	63.25±* 0.53
Oxygen consumpt-ion (ml/min)	272.06± 1.56	267.7± 1.8	297.41± 2.47	294.7±* 2.93	322.1±* 1.62	321.5±* 1.82	349.5±* 2.03	354.9±* 2.02
Oral temperature (°C)	36.97± 0.02	36.96± 0.02	36.53± 0.06	36.37±* 0.03	36.26±* 0.05	36.05±* 0.03	36.03±* 0.037	35.95±* 0.03
Mean skin temperature (°C)	30.57± 0.31	30.66± 0.24	29.25± 0.25	29.32± 0.23	27.58±* 0.29	26.56±* 0.18	26.04±* 0.26	25.48±* 0.19
Visible shivering	Nil	Nil	Disconti-nuous mild	Mild discon-tinuous	Moderate	Moderate	Continuous and severe	Continuous and severe

* $p < 0.01$

DISCUSSION

This study has demonstrated that acute exposure to cold is associated with a marked rise in plasma T_4 and T_3 levels. This rise in T_4 and T_3 could be recorded four hours after the cold exposure and the elevated levels were maintained till the duration of cold exposure schedule. Contrary to the observations in plasma, T_3 concentration of erythrocytes was markedly decreased on day 1 of exposure and returned to preexposure levels on day 6. TSH values in the plasma did not show any significant change on day 1 but showed a decline on day 6.

Earlier studies on cold exposure in man have reported many metabolic adjustments. Cold exposure has been shown to result in an enhanced adrenocortical, adrenomedullary and sympathetic activity (Suzuki et al., 1966; Wilson et al., 1970), increased metabolic rate and oxygen consumption (Mathew et al., 1981) and decrease in plasma volume (Wilson et al., 1970). These changes are likely to have a significant bearing on the thyroid hormone economy. An increase in T_4 secretion rate and turnover rate has been recorded in experimental animals (Bauman and Turner, 1967; Hillier, 1968) and men (Bass, 1960; Ingbar and Bass, 1967) exposed to a cold environment. A similar mechanism might be operating in the case of T_3 also. However, the observation that T_3 concentration of erythrocytes was markedly decreased on day 1 of exposure suggests that an increase in T_3 levels in the plasma on day 1 could also be contributed by a shift of hormones from intracellular to extracellular fluid compartment. Although we could not measure T_4 concentration of RBCs, a similar explanation can also be extended to its observed rise on day 1 of exposure.

Haemoconcentration and increase in total proteins reported during cold exposure (Wilson et al., 1970) may also vitiate the measurements of hormones in the plasma. However, the magnitude of increase in serum proteins and haematocrit of 6 to 11% reported by different investigators (Bass and Henschel, 1956; Suzuki et al., 1966; Wilson et al., 1970) is not sufficient to explain the much greater increase of 91% and 167% in respect of T_4 and T_3 recorded in our study. Moreover, since serum osmolarity has also been reported as normal in men subjected to cold exposure (Eastman et al., 1974), and measurements of T_4 and T_3 in serial dilutions of plasma obtained after cold exposure did not reveal any unexpected changes in their assayed values, the observed increase in T_4 and T_3 after cold exposure cannot be attributed to the above mentioned factors. Increased T_4 and T_3 levels are unlikely to be secondary to rise in thyroid hormone binding proteins, since TBG levels (Rastogi and Sawhney, 1976) and ^{131}I-T_3 resin uptake (Sukuzi et al., 1966) has been reported as normal in cold-exposed men.

Animals exposed to cold have been shown to exhibit a rapid increase in TSH secretion within the first few hours of exposure (Bottari, 1957; Itoh et al., 1966). However, in man, no change in TSH levels on day 1 of exposure was observed in the present study. Our observations on unaltered TSH levels on day 1 of cold exposure are in agreement with those of Wilson et al. (1970), Hedstrand and Wide

(1973) and Rastogi and Sawhney (1976), but do not conform with the findings of Suzuki et al. (1966) and those of Raud and Odell (1969) who observed an increase in TSH levels following cold exposure in man. Fisher and Odell (1971) have reported TSH increase after cold in infants but failed to demonstrate any such changes when adult men were subjected to cold exposure. Prolongation of the cold exposure for six days in our study resulted in a further increase in plasma T_3 and normalization of T_3 concentration of erythrocytes. This increase in T_3 levels in the plasma probably resulted in decreased thyrotropin secretion from the anterior pituitary.

In spite of elevated T_4 and T_3 levels in the plasma, the physiological responses to cold remained unaltered. Because of the long latent period of thyroid hormone action, the thyroid may have no direct involvement with the initial phase of acclimatization to cold, though it may be important in potentiating the catecholamine actions (Suzuki et al., 1966).

ACKNOWLEDGEMENTS

The authors are grateful to Group Capt K.C. Sinha, Director, Defence Institute of Physiology and Allied Sciences, for his encouragement and keen interest in this study. Secretarial assistance of Mrs Anita Sethi is also acknowledged.

REFERENCES

BASS, D.E. (1960): Metabolic and energy balances in men in a cold environment. Transactions of the 6th Conference of cold injury, ed. Horvath S.M., pp. 317-338. Josiah Macy Jr. Foundation, New York.

BASS, D.E. and HENSCHEL, A. (1956): Responses of body fluid compartments to heat and cold. Physiol. Rev., 36: 128-144.

BOTTARI, P.M. (1957): The concentration of thyrotropic hormone in the blood of rabbit under different experimental conditions. Ciba Found. Colloq. Endocrinol., 11: 52-69.

BAUMAN, T.R. and TURNER, C.W. (1967): The effect of varying temperatures on thyroid activity and the survival of rats exposed to cold and treated with thyroxine or corticosterone. J. Endocr., 37: 355-359.

COTTLE, W.H. and CARLSON, L.D. (1956): Turnover of thyroid hormones in cold exposed rats determined by radioactive iodine studies. Endocrinology, 59: 1-11.

EASTMAN, C.J., EKINS, R.P., LEITH, I.M. and WILLIAMS, E.S. (1974): Thyroid hormone response to prolonged cold exposure in man. J. Physiol., 241: 175-181.

FISHER, D.A. and ODELL, W.D. (1971): Effect of cold on TSH secretion in man. J. Clin. Endocr. Metab., 33: 859-862.

HEDSTRAND, S. and WIDE, L. (1973): Serum thyrotropin levels in summer and winter. Br. Med. J., 4:420.

HILLIER, A.P. (1968): Thyroxine deiodination during cold exposure in the rat. J. Physiol., 197: 135-147.

INGBAR, S.H. and BASS, D.E. (1967): The effect of prolonged exposure to cold on production and degradation of thyroid hormone in man. J. Endocr., 37: II-III.
ITOH, S., HIROSHIGE, T., KOSEKI, T. and NAKATSUGAWA, T. (1966): Release of thyrotropin in relation to cold exposure. Fed. Proc., 25: 1187-1192.
LEBLOND, C.P. and EARTLY, H. (1952): An attempt to produce complete thyroxine deficiency in the rat. Endocrinology, 51: 26-41.
MATHEW, L., PURKAYASTHA, S.S., JAYASHANKAR, A. and NAYAR, H.S. (1981): Physiological characteristics of cold acclimatization in man. Int. J. Biometeor., 25: 191-198.
RASTOGI, G.K. and SAWHNEY, R.C. (1976): Thyroid function in changing weather in a subtropical region. Metabolism, 25: 903-908.
RASTOGI, G.K., SINHA, M.K., DASH, R.J. and KANNAN, V. (1973): Plasma thyroid stimulating hormone (TSH) levels in health and thyroid disorders. J. Assoc. Phys. Ind., 21: 183-188.
RAUD, H.R. and ODELL, W.D. (1969): The radioimmunoassay of human thyrotropin. Brit. J. Hosp. Med., 2: 1366-1376.
SUZUKI, M., TOMOUE, T., MALSUZAKI, S. and YAMAMOTO, K. (1966): Initial response of human thyroid, adrenal cortex, and adrenal medulla to acute cold exposure. Canad. J. of Physiol. Pharmacol., 45: 423-432.
WILSON, O., HEDNER, P., LAURELL, S., MOSSLIN, B., RERUP, C. and ROSENGREN, E. (1970): Thyroid and adrenal response to acute cold exposure in man. J. Appl. Physiol., 28: 543-548.
WOODS, R. and CARLSON, L.D. (1956): Thyroxine secretion in rats exposed to cold. Endocrinology, 59: 323-330.
YOSHIDA, K. and DAVIS, P.J. (1980): Estimation of intracellular free triiodotyronine in man. J. Clin. Endocr. Metab., 50: 667-669.

EFFECT OF MAGNETIC MICROPULSATIONS ON THE BIOLOGICAL SYSTEMS - A BIOENVIRONMENTAL STUDY

Sarada Subrahmanyam, P.V. Sanker Narayan
and T.M. Srinivasan
(Voluntary Health Services Medical Centre,
Adyar, Madras - 600 113)

Abstract: - During the last decade considerable interest has been evinced by scientists on the possible influence of the earth's electromagnetic environment on human and animal physiology. While some studies on this topic have been reported from high magnetic latitudes - USSR and central Europe - no work has been done in very low latitude and equatorial regions. The present study, undertaken to fill this gap, has been carried out at the low latitude of Madras (Magnetic Dip $\cong 10°$). Pulsating magnetic fields in the frequency range of 0.01 Hz to 20 Hz and with amplitudes of ± 5 and ± 50 nano Teslas (nT) were impressed on test animals, normal human subjects and Yoga practitioners lying supine inside a 4-member Fanselau-Braunbeck coil system with the heads oriented in the four cardinal directions with respect to earth's magnetic field. The entire set of exposure of the test animals and humans was given under two ambient magnetic fields namely, against full local geomagnetic field of about 40,000 nT and half this value. In the animals ECG, EEG, tail blood flow and respiration were recorded continuously on a polygraph. The biochemical tests carried out were postprandial blood sugar, serum cholesterol and plasma cortisol. Neurochemical assays of Noradrenaline, Adrenaline, Dopamine, Serotonin and 5-Hydroxy Indole Acetic Acid were done in the brain tissue, myocardium and adrenal glands, immediately after complete set of exposure of the animals in all four orientations. Motor activity and rectal temperature were also noted before and after the exposures. The 'Control' animals were subjected to exactly the same investigations as the test animals without, however, exposing them to the magnetic fields. These observations revealed some dicisive changes in certain parameters for certain frequencies of the impressed field and also in specific orientations of the test animals. Similar studies carried out on normal human subjects and practitioners of Yoga and Meditation also showed certain decisive changes in the electrophysiology, neurochemistry and biochemistry when oriented to North and East. The North orientation appeared to induce inhibition of brain electrical activity and associated neurochemical and biochemical changes, whereas the East orientation showed a response of calm, blissful alertness.

INTRODUCTION

One of the earliest and most compelling promptings from the knowledge explosion that space sciences witnessed during the last two decades was the awareness of the possible control of the space environmental fields on our biosphere. Possibly stemming from the anxiety as to how man would withstand and survive in the rather alien environment of outer space during future space colonization programmes, medical and biological scientists started examining the problem on the earth's surface itself.

Soviet scientists were the earliest to take up investigations on the possible influence of the earth's electromagnetic field on our biosphere - in particular on the physiology of man. It has been pointed out in very recent times that large-scale changes in the earth's magnetic field or other abrupt changes of the geomagnetic field intensity are directly correlatable with recorded major evolutionary catastrophies of the geologic past - such as abrupt and total extinction of species or major changes in the skeletal structure of faune etc. (Dubrov, 1978) and the conviction has grown that these are not to be taken as casual coincidences; rather, that they are expressions of major "cause and effect" episodes in nature's perpetual homeostatic war with the environment.

It is a hardly appreciated fact that we are enveloped from our embryonic stage of existence by an electromagnetic field which is strong, complex, extremely wide-spectrummed and continually changing, both on long-term basis as well as on micro-time scale. Adding to this our unprecedented technological developments over the last few decades have swamped our other environment with electromagnetic radiations of extremely low to super-high frequencies generated by the thousands of radio and T.V. stations, high tension power transmission lines, industries and domestic appliances like microwave ovens and the like. The ill-effects of pollution by super-high frequency electromagnetic fields has already attracted the attention of WHO and exclusive expert committees are looking into the safe levels or dosages for man so that the so called 'pollution levels' for these could be defined.

The other end of our spectrum is formed by the Extremely Low Frequency (ELF) electromagnetic fields which are generally known to be essentially energised by natural pulsations of the earth's magnetic field, electric discharges of thunderstorms etc. and so far believed to be comparatively inconsequential to human existence. However, this by no means gives us a scientifically tenable assurance until so proved. In fact to keep an absolutely open, scientific mind to the question, it should be examined whether these ELF fields have any effects - ill or beneficial - on our biosphere in general; if ill effects, to find out ways of shielding, minimising or even eliminating such effects and, if beneficial, to see how they could be used in a controlled fashion, as possible therapeutic agents in the alleviation of human suffering.

With intentions such as these a small group has been working since the last two years on the effects of ELF magnetic fields on test animals, healthy humans and Yoga practitioners under a Department of Science and Technology supported scheme at the Voluntary Health Services Medical Centre, Adyar, Madras.

The first phase of our investigations, carried out over the last two years, covered experiments on test animals (Sarada Subrahmanyam et al., 1983), Since the last year we have extended the experiments to normal human subjects and Yoga practitioners as well (Sarada Subrahmanyam et al., 1983). A fact of observation faintly suspected by some workers in this line is that the effects of EM fields on biological objects may be confined to certain discrete 'windows' of the spectrum. That is, the fields of certain band of frequencies and certain intensities (not necessarily high) seem to affect biological objects. Similarly it has been hypothesized that the effects might possibly be impressionable only against certain ambient steady (magnetic) fields. The complexity of any investigations in this realm can be appreciated from these parametric backgrounds.

EXPERIMENTALS

For experiments on human subjects a 4-member Fanselau-Braunbeck coil system with the smaller and larger coil pairs of radii 148.5 and 194.4 cm and with a Steady Field Winding of 12 turns and a Pulsation Winding of 3 turns of SWG-20 superenamelled copper wire was fabricated (Sanker Narayan et al., 1982). The four coils were mounted co-axially with the axis along the magnetic meridian and dipping 10° from horizontal towards North (this being the magnetic Inclination for Madras) so that any magnetic field generated inside the coil system by currents flowing in the four coils in series, will vectorially add or oppose the local Total Magnetic Field of about 40,000 nT. The enclosure covers a floor space of about 400 x 350 cm and reaches upto 415 cm above the floor (Fig. 1).

An identical 4-member coil system of much smaller dimensions (smaller and larger coil radii being 38 cm and 50 cm resp.) with a Steady Field Winding of 10 turns and Pulsation Winding of 3 turns) was fabricated for experiments with test animals. The pulsation winding, energized by a signal generator can maintain within the coil system, pulsating magnetic fields of any frequency, amplitude and waveform such as sine, triangular, square or saw tooth. By sending direct currents of appropriate value and direction through the Steady Field Windings the ambient magnetic field inside the coil system can be maintained at any value (direction remaining the same as that of the local Total Magnetic Intensity) or even practically zero.

MATERIALS AND METHODS

Studies on animals:

In the experiments on animals male albino-rats (120-150 g weight) partly restrained in a perspex harness and held at the centre of the C.M.F. coil system were exposed to sinusoidal, pulsating magnetic

Figure 1: *Controlled Magnetic Field Enclosure and redording setup.*

fields of frequency 0.01, 0.1, 1.10 and 20Hz and amplitudes of ±5 and ±50 nT with their heads facing magnetic North, East, South and West. The experimental animals were first 'preconditioned' for the exposure by being held in the enclosures with the electrode assemblies in position but without any pulsations being impressed on them for as long as two hours or more and were subjected to actual exposure only after the levels of catecholamines and/or plasma cortisol became normal these being taken as an indication of freedom from fright stress. Besides the experimental animals, another identical set of 'control' animals were also placed separately in the same enclosure without, however, being exposed to the pulsating fields.

During the exposures ECG, EEG, Tail Blood Flow and Respiration were continuously recorded on a polygraph. The experimental animals were sacrificed after exposure in each frequency and orientation, the brain removed and the biogenic amines NA, A, DA, 5-HT and 5-HIAA assayed from different regions of the brain as also from total brain. Blood was collected by cardiac puncture for biochemical and haematological studies. In addition, adrenal glands and heart were taken out and the amines assayed. The biochemical investigations covered sugar, cholesterol and plasma cortisol in blood and MHPG, HVA, VMA, 5-HIAA, Catecholamines and 17-ketosteroids in urine, by standard methods.

Studies on human subjects:

The subjects (normal human subjects and practitioners of Yoga and meditation) lying supine on a couch at the centre of the CMF coil system with the head towards the four cardinal directions with respect to the direction of the local magnetic field (north, east, south and west) and also in sitting posture, were exposed to pulsating magnetic fields of frequencies 0.01, 0.1, 1.0 and 10 Hz and amplitudes ±5 and ±50 nT in ambient geomagnetic field (about 40,000 nT at the experiment location) and also half this value. The subjects were 'pre-conditioned' for the exposures by keeping them in the enclosures with the electrode assembly in position for a few hours and also at least two hours prior to exposure in each case and were subjected to the final exposures only after the levels of catecholamines and/or plasma cortisol became normal, this being taken as an indication of freedom from fright/anxiety stress. Prior to the experiments blood and urine samples were collected on the previous day and also on the day of experiment before and after exposure in each frequency and amplitude, to serve as control data.

The following environmental conditions were carefully maintained/observed in each case:

(a) The recording equipment as also the field generating devices together with the members operating these were all isolated from the CMF enclosure room to ensure that the subjects were not aware of the setup or procedural stages of the experiment. However, the operators could see the subjects through a special observation window throughout the various stages of the experiment.

(b) The actual times of onset and cessation of the fields - steady and pulsating - were not known to the subjects.

(c) The entire CMF room was kept as noise-free as possible; in particular no conversations/instructions between the operators were audible to the subjects.
(d) The subjects were not aware of the absolute orientations in which they were placed.
(e) It was ensured by reference to the Geomagnetic Observatory of the National Geophysical Research Institute, Hyderabad about 500 km to the North-North-West of Madras and the Micropulsation Observatory at Etaiyapuram about 500 km to the South-South-West of Madras that there was no major or moderate geomagnetic disturbance (including pulsation activity) during any of the experiments.

Before and during the exposures in all orientations ECG, EEG, Finger Blood Flow and Respiration were continuously recorded on a polygraph. Also, during the exposures to the impressed pulsating field, EEG was recorded on an 8-channel EEG machine from different regions of the brain and EEG from one channel was split up into the alpha, beta, theta and delta bands by four filters. ECG was also recorded separately in the classical limb and chest leads. The subjective experiences of each subject in each orientation were also elicited.

Two days before the experiments were started, blood and urine samples were collected from all subjects and again on each day after exposure to each frequency and amplitude. In the blood, sugar, cholesterol, plasma cortisol and serum-protein bound iodine were estimated and in the urine, MHPG, VMA, Total CA, HVA and 5-HIAA were assayed as also 17-Ketosteroids, using conventional methods.

RESULTS

The experiments on animals showed certain decisive changes in the electrophysiological, neurochemical and biochemical parameters. These will be outlined briefly so as to provide a link with the more interesting results observed with humans.

ANIMAL STUDIES

Electrophysiological: Marked changes are noticed in the cardiac functions during exposure to 0.01 and 0.1 Hz frequency pulsations, especially in the North orientation of the animals, other orientations and frequencies hardly showing any significant changes. The ECG changes are suggestive of the impulses originating in an ectopic focus (Fig. 2). The animals appeared to be restless and aggressive when exposed to pulsations in the North orientation and screeched frequently.

In the east orientations the animals do not show any changes in their ECG, EEG, Respiration and Tail Blood Flow.

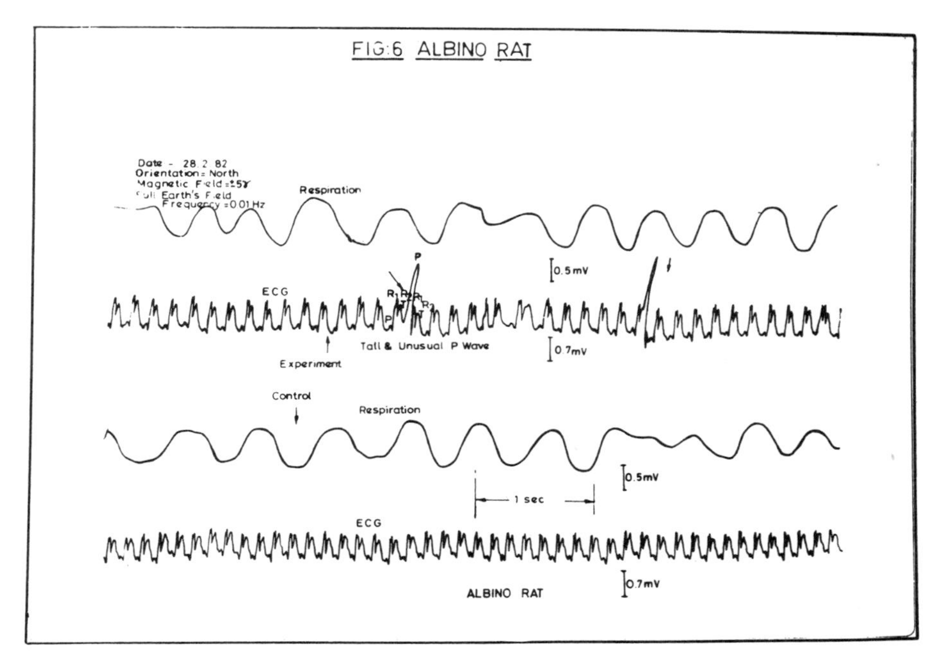

Figure 2: *Polygraphic recording of ECG and Respiration in a 'Control' rat and one exposed in North orientation.*

Neurochemical: In the total brain and in the different regions the NA and DA levels were increased while 5-HT and 5-HIAA diminished on exposure in the North specially to frequencies of 0.01 and 0.1 Hz whereas hardly any effect was noticeable in the East orientation (Table 1).

Biochemical: After exposure in the North orientation against full geomagnetic ambient field the blood sugar showed an increase whereas there was no significant changes in cholesterol and plasma cortisol.

HUMAN STUDIES

The EEG of a normal subject before and after exposure to 0.01 Hz pulsations with amplitude of ±50 nT with the head towards East is depicted in Fig. 3. There was accentuation of alpha and beta rhythms indicating a state of restful mental alertness. In the Yoga practitioners the accentuations of the waves was more prominent than in normal subjects. The subjects felt calm and relaxed and expressed a state of blissfully pleasant feelings. Maximum effects were noticed in the frequencies of 0.01 and 0.1 Hz and in both the amplitudes (±5 and ±50 nT).

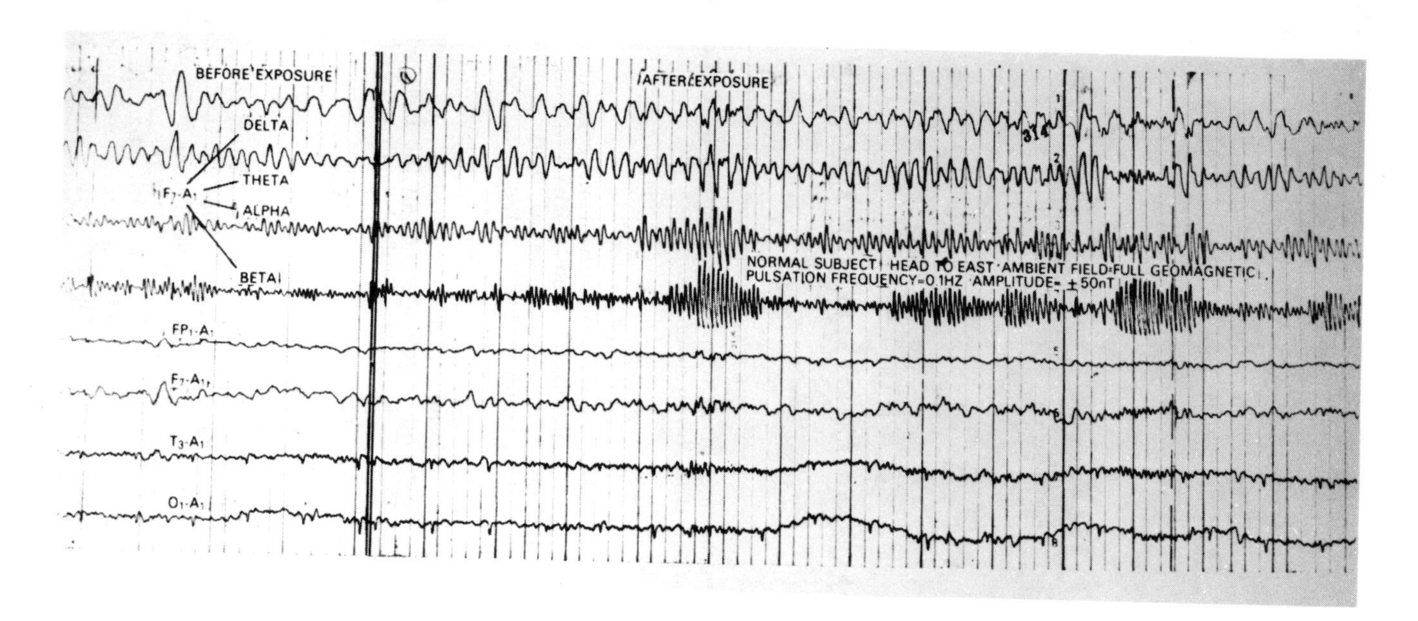

Figure 3: *EEG of a normal human subject in East orientation with 0.1 Hz, ± 50 nT pulsations.*

TABLE 1: *Rat brain - full earth's field biogenic amines.*

Orientation Animal facing	No.	µg/g of wet tissue				
		Na	A	DA	5-HT	5-HIAA
North	10	4.177 ± 0.105	1.118 ± 0.073	10.514 ± 0.120	1.141 ± 0.083	3.367 ± 0.121
South	10	3.842 ± 0.185	1.160 ± 0.052	10.492 ± 0.110	1.139 ± 0.075	3.108 ± 0.101
East	10	3.210 ± 0.173	1.174 ± 0.065	9.890 ± 0.120	1.326 ± 0.081	4.123 ± 0.120
West	10	3.180 ± 0.168	1.820 ± 0.073	9.912 ± 0.131	1.368 ± 0.076	4.360 ± 0.118
Control	10	3.151 ± 0.132	1.193 ± 0.078	9.927 ± 0.126	1.376 ± 0.078	4.427 ± 0.112

Values are mean ± S.E.M.

Fig. 4 depicts the polygraphic recording of Respiration, Finger Blood Flow and ECG before and after exposure to pulsating field of the same frequency and amplitude and shows an increase in the peripheral blood flow. No significant changes were seen in the Respiration and ECG.

In normal subjects with the head oriented to the North the electrical activity of the brain in general was reduced considerably as compared with that recorded before exposure (Fig. 5). Even in Yoga practitioners there was considerable reduction in the electrical activity of the brain when in North orientation. The subjects experienced uneasiness, confusion of mind and, in general, restlessness. The conjunctiva became red and they complained of headache after coming out of the field for a few hours. The effects were less severe in Yoga practitioners. However, a careful ophthalmological examination of the fundus did not reveal any change.

In the polygraphic recording with North orientation the peripheral blood flow was decreased in normal subjects as well as in Yoga practitioners (Fig. 6) with no change in east orientation.

Certain neurochemical and biochemical changes were also noticed in the North orientation. In the blood, sugar and cortisol were increased when in the North orientation whereas no significant changes were noticed in all the other orientations with normal subjects (Table 2). In Yoga practitioners no significant changes were noticed in any of the parameters.

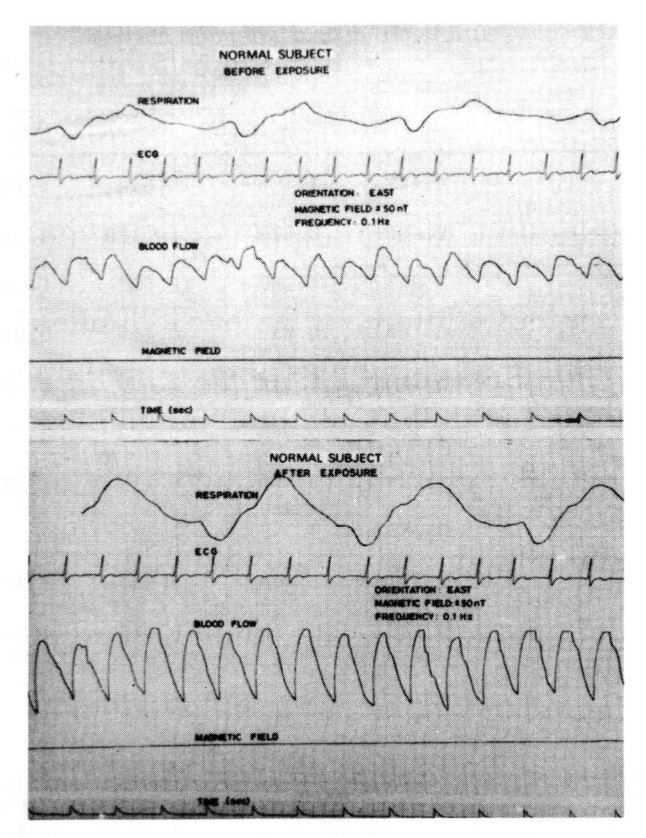

Figure 4: *ECG, Finger Blood Flow and Respiration of a normal human subject in East Orientation with 0.1 Hz, ±50 nT pulsations.*

In the urine, the metabolite of serotonin was reduced both in the normal subjects and Yoga practitioners - HVA, the metabolite of Dopamine increased. 17-Ketosteroids also showed an increase in North orientation (Table 3).

DISCUSSION

Examining the results of the experiments done so far on both animals and man, it appears at our very low magnetic latitudes (close to the magnetic equator) that the North orientation - that is, with the body horizontal and head towards North - does generate profound changes in the electrophysiology and neurochemistry and biochemistry of the

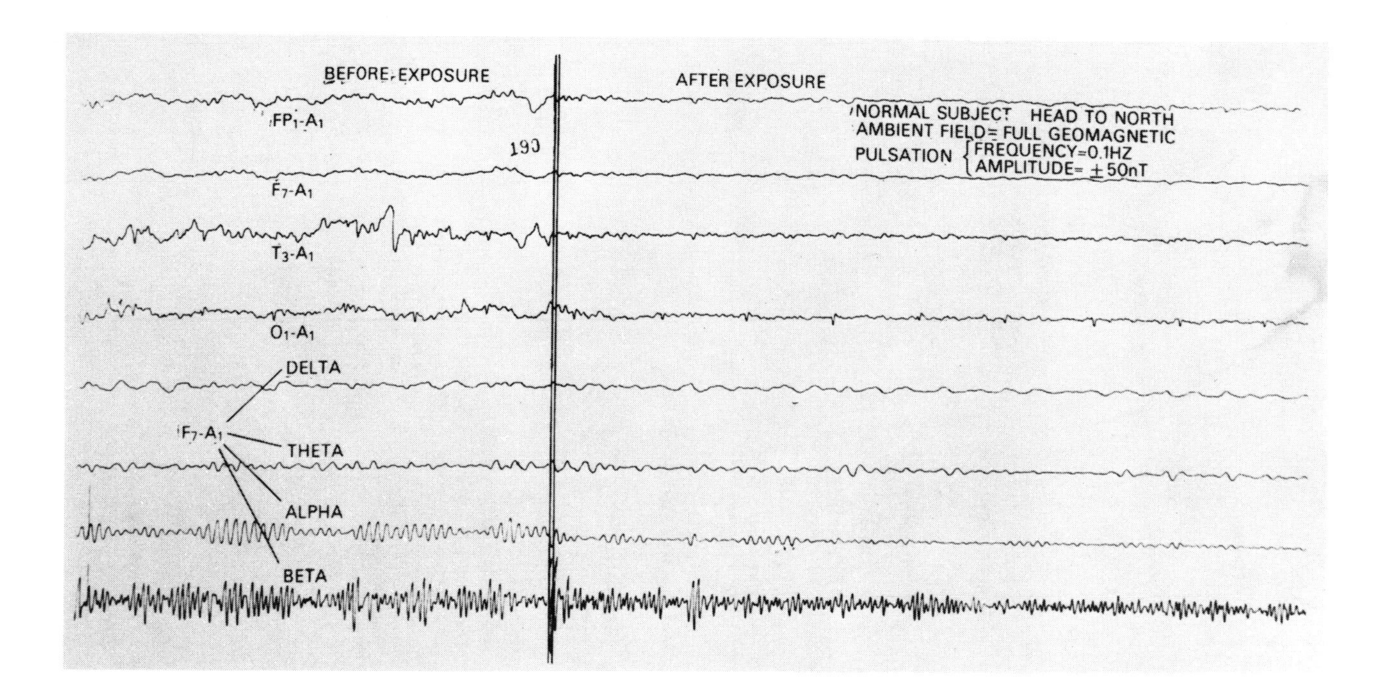

Figure 5: *EEG of a normal human subject in North orientation with 0.1 Hz, ± 50 nT pulsations.*

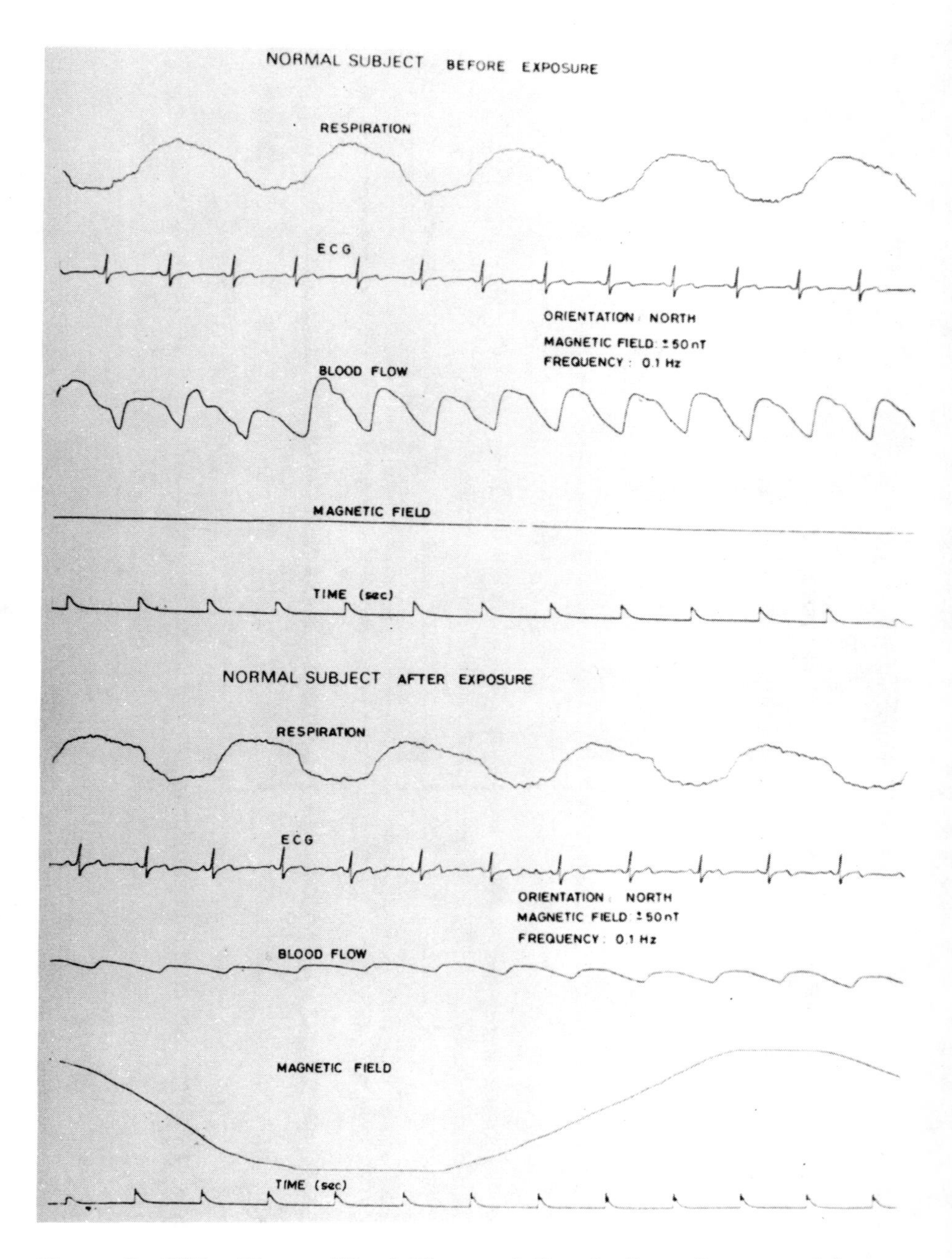

Figure 6: *ECG, Finger Blood Flow and Respiration of a normal human subject in North orientation with 0.1 Hz, ± 50 nT pulsations.*

TABLE 2: *Blood-pulsation frequency 0.01 Hz. Amplitude ± 50 nT field - full geomagnetic.*

Orientation	Group	Sugar	Chole-sterol	P.B.I.	Plasma-cortisol
		mg%	mg%	μg%	μg%
Control (North)	Normal	96 ± 4.0	185 ± 10.2	5.00 ± 1.00	9.8 ± 1.1
East	"	96 ± 6.2	185 ± 8.4	5.00 ± 0.80	9.6 ± 0.82
West	"	98 ± 4.6	185 ± 8.6	5.10 ± 1.2	9.8 ± 1.2
South	"	98 ± 4.2	190 ± 7.2	5.20 ± 1.4	9.4 ± 1.6
North	"	104 ± 6.0	190 ± 8.8	5.80 ± 1.0	11.1 ± 1.0
North	Yoga	95 ± 2.2	188 ± 6.6	5.30 ± 0.82	10.1 ± 1.4
Sitting	Normal	98 ± 4.8	185 ± 5.0	5.20 ± 0.60	9.4 ± 1.1

Values are mean ± S.E.M.

TABLE 3: *Urine-pulsation frequency 0.01 Hz. Amplitude ± 5 nT full geomagnetic field.*

Orientation	Group	5-HIAA	HVA	17-Keto-steroids
		mg%	mg%	mg%
Control (North)	Normal	0.39 ± 0.01	0.35 ± 0.02	0.90 ± 0.10
East	"	0.36 ± 0.08	0.35 ± 0.03	0.98 ± 0.16
West	"	0.36 ± 0.06	0.37 ± 0.08	1.04 ± 0.08
South	"	0.36 ± 0.04	0.38 ± 0.04	1.08 ± 0.06
North	"	0.29 ± 0.02	0.44 ± 0.04	1.20 ± 0.10
"	Yoga	0.28 ± 0.02	0.28 ± 0.02	1.06 ± 0.06
Sitting	Normal	0.34 ± 0.08	0.40 ± 0.06	0.96 ± 0.04

Values are mean ± S.E.M.

subjects, when exposed to pulsating magnetic fields of frequency around 0.01 to 0.1 Hz and amplitudes in the range of a few to a few tens of nT. These pulsations almost exactly simulate the natural pulsations of the earth's magnetic field - the so-called Pc_1, Pc_2, Pc_3, Pc_4, and Pc_5 class as also the Pi_1, though their amplitudes at our latitudes rarely go to tens of nT. The animals, under the above conditions, show great discomfort and aggressiveness. There was an upset in the balance between the catechol and indole amines in North orientation. The same results are seen whether the ambient field is the full local geomagnetic field or half its value.

As opposed to this, when in the East orientations the animals show hardly any changes in their brain activity, biochemistry etc. In the other two directions, South and West, there were only minor changes.

In the case of humans very similar results are seen. In the North orientation with pulsations of frequencies 0.01 and 0.1 Hz and amplitudes of ±5 and ±50 nT there is considerable inhibition of brain's electrical activity as evinced by the almost abrupt and strong dampening of all the four rhythms - in particular the alpha wave. As to the subjective experiences it has been indicated that they all had confusion of thinking, mental irritation and a lack of sense of well-being when subjected to the pulsating fields in North orientation. As opposed to this when exposed to the same pulsating field in the East orientation the subjects experienced a calm and blissful feeling.

In the case of practitioners of Yoga and meditation the above effects are much less pronounced. This seems to be in conformity with the generally accepted and now proven observation that a Yoga-trained person is able to adjust the balance between the levels of the various neurotransmitters at the most desirable state even when subjected to stress-generating fields or environment.

We are cautious about offering any formal and rigorous interpretation of the findings reported here since (a) we feel a lot more experimenting needs to be carried out to the point of total repeatability before the results can be considered conclusive, and (b) we do not yet know enough about the various levels at which an electromagnetic stress field of given specifications interacts with the various systems of the human body - at ionic levels across cell membrane, or neurotransmitters, or on bulk electrical characteristics of body fluids or on the generation and propagation of action potentials in the central nervous system or even on the recently-discovered ultra weak, primary magnetic and electric fields in the brain.

ACKNOWLEDGEMENT

We are thankful to the Department of Science and Technology for the financial assistance for the research project and also to the Secretary, Voluntary Health Services Medical Centre, Adyar, Madras for the various facilities provided.

REFERENCES

DUBROV, A.P. (1978): "The Geomagnetic Field and Life", Translated from original Russian by F.L. Sinclair, Plenum Press, New York, Chapter 3.

SUBRAHMANYAM, S. and SANKER NARAYAN, P.V. (In press): Effect of pulsating magnetic fields on albino-rats. Proc. XXIX International Congress of International Union of Physiological Sciences, Sydney, Australia, 1983.

SUBRAHMANYAM, S., SANKER NARAYAN, P.V. and SRINIVASAN, T.M. (In press): Effect of magnetic micropulsations on biogenic amine levels in animals and in human subjects. Proc. III Symp. of Catecholamines and other Neurotransmitters in Stress. Czechoslovakia, 1983.

SUBRAHMANYAM, S. and SANKER NARAYAN, P.V. (In press): Physiological effects of pulsating electromagnetic fields. Proc. Ist Soviet-Indian Symposium "Neurophysiology", Tzahkadzor, USSR, 1983.

SANKER NARAYAN, P.V., SUBRAHMANYAM, S., SRINIVASAN, T.M., MURUGAN, S., and PETRAITIS, F. (1982): A controlled Magnetic Field (CMF) Enclosure for Experiments in Magnetophysiology, J. Biomed. 2: 25-29.

WEATHER AS THE CAUSE OF HUMAN AILMENTS AND ITS ROLE IN THE SELECTION OF HOMOEOPATHIC REMEDIES

P.K. Misra
(Director, Area Cyclone Warning Centre,
Regional Meteorological Centre,
Colaba, Bombay-400 005)

Abstract: - A certain biological effect, as a result of specific weather changes, often does not develop in every individual of a population at the same time and to the same extent. The genetic factors of an individual basically determine the pattern of his reaction to environmental changes. Hence, in the homoeopathic branch of medicine, the constitution of an individual and the climatic factors have been given due consideration in the selection of remedies.

INTRODUCTION

Several research studies have been published on the influence of weather and climate on human diseases, most of these correlations are statistical in nature, but in a few instances the investigators have elucidated the deeper physiological mechanisms involved. A certain biological effect, as a result of specific weather changes, often does not develop in every individual of a population at the same time and to the same extent. This is due primarily to the fact that every person has a different genetic background. The genetic factors of an individual basically determine the pattern of his reaction to environmental changes. In other words, the constitution of an individual basically determines his susceptibility to sudden environmental changes. In homoeopathy, even much before the discovery of the genetic concept, the constitution of an individual had been given a foremost place in the selection of remedies. The basic philosophy of the homoeopathic branch of medicine is "treat the patient, not the disease".

The healthy functioning of the human organism is closely dependent upon the physical and chemical properties of the internal environment. The properties of the internal environment exhibit limited or regulated variability. Therefore the maintenance of good health essentially means maintenance of the "steady state" within the internal environment. The internal and external regulatory mechanism of the human body make it possible to maintain a steady state within the internal environment. Whenever these regulatory mechanisms fail to maintain the steady state within permissible limits, an individual becomes sick.

THERMAL IMBALANCE - HEAT LOAD

The temperature of the internal environment of a human body is close to 37° C. The thermoregulatory mechanism inside the human body functions in such a manner that it balances heat production and heat loss so that the temperature of the internal environment is preserved. Human heat balance consists of gains from metabolism and absorption of solar and infra-red (IR) radiations and of losses from outgoing IR, convection and evaporation. The strain resulting from radiation heat load on the body is always more relevant because man cannot so easily protect himself from this heat load especially when he is outdoors. Prolonged exposure to high sun in summer may lead to heat strokes.

An overdose of solar ultra-violet (UV) leads to severe erythema, oedema and blistering. Frequent exposure to UV also causes skin ageing in the form of wrinkling and loss of skin texture. Finally these may lead to skin cancer which is restricted to the frequently exposed areas like face, neck and limbs. Many sun-synchronous headaches and skin diseases which evade cure from other branches of medicines readily respond to weather-oriented homoeopathic remedies. Some of the hot weather remedies are discussed below: -

Glonoinum:

The most common symptoms which are generally observed in cases of sun-strokes get aleviated by glonoinum. It is also equally indicated for the after-sufferings from sun-stroke. Glonoinum is also indicated in ailments which result from other bad effects of radiated heat; for instance, children get sick in the night after sitting too long or falling asleep before an open coal fire.

Antimonium Crudum:

Antimonium Crudum is another remedy which takes a high rank as a hot weather remedy. The conditions which need this remedy become particularly worse in sun-shine as well as from radiated heat of the fire. A child with whooping cough will cough more after looking into the fire. Skin diseases like skin cancer, eczema, boils etc. which are mainly restricted over those parts of the body which are exposed to the sun, are often cured by this remedy.

A typical case history:

A young boy developed a peculiar type of skin disease. All parts of his body which was exposed to the sun became like a frog's skin. Eruptions over the exposed areas used to become bold whenever he went out into the sun. After being disappointed by many eminent dermatologists of Bombay, the father of the boy approached a novice and amateurish homoeopathic practitioner, who cured the boy in first stroke by prescribing him Antim Crudum.

Spigelia Anthelmintica:

The headaches which are generally one-sided increase with the rising of the sun and decrease with its going down.

Persons suffering from overheating from the sun may also find relief from remedies like *natrum carbonicum, Lachesis* or *Lyssin.*

THERMAL IMBALANCE - COLD

The sensation of cold gives rise to behavioural adaptation to cold. One line of defence is to seek more clothing or a warm environment. The other line of defence is to increase the internal heat production, either by increased muscle tone and shivering or by non-shivering thermogenesis (heat production by chemical processes independent of muscular action). In the case of people highly acclimatized to cold, non-shivering thermogenesis adequately supplies heat at low temperatures whereas non-acclimatized people shiver violently so that extra heat is produced by the burning of carbohydrate.

In homoeopathy symptoms like chilliness, sweat, shivering, likes and dislikes for coverings etc. which are intimately connected with the process of heat loss from a human body, play a great role in the selection of remedies. While selecting a remedy, a clear distinction is necessary to be made whether the patient is more susceptible to dry cold air/environment or to wet cold air/environment. The development of these symptoms is basically connected with the genetic state of the individual.

SOME OF THE REMEDIES WHICH SHOW AGGRAVATION FROM DRY COLD WEATHER

Aconitum Napellus:

This remedy is suitable for complaints which result from exposure to dry cold air, in particular to dry cold northwesterlies. Patients who are plethoric and vigorous have a strong heart and vigorous circulation. When exposed to dry cold weather they come down very rapidly with violent symptoms. Like a great storm their diseases come, sweep over and pass away. Diseases like croup, bronchitis, pneumonia, rheumatism, inflamatory fever which are caused by exposure to cold north or westerly winds which prevail over north India immediately after the passage of an active western disturbance, respond very well to this remedy if given in the initial stage. Though the complaints are caused due to dry cold air, the patient removes covering during the stage of high inflamatory fever. An Aconite patient does not sweat at all.

Hepar Sulphuris:

The strongest characteristic of this remedy is its hypersensitiveness to touch, pain and dry cold air. The patient starts coughing when any part of the body becomes uncovered. This is found in croup, laryngitis, bronchitis and consumption, and not only the cough is worse but the whole case is aggravated from dry cold wind. The chronic asthma, which is worse in dry cold air but is greatly ameliorated in damp weather, is cured by this remedy. The bronchial asthma which does not show aggravation with the arrival of a western disturbance but is greatly aggravated immediately after the passage of the active disturbance, may be cured by this remedy provided other symptoms agree. Hepar is very useful in chronic catarrh when the nose gets blocked everytime the patient goes out into the cold air. It is relieved in a warm room. The Hepar patient also sweats day and night without relief during his complaints.

Kali Carbonicum:

The Kali Carb patient is also very sensitive to cold and is always shivering. He is sensitive to every draft of air and to the circulation of the air in the room. He cannot have the windows open, even in a distant part of the house. His complaints are worse in cold air - whether dry or moist. Kali Carb patients keep well in warm weather, even during wet weather provided the air is warm. Asthmatic patients whose complaints are aggravated in cold weather - both dry or wet - but are ameliorated in warm weather including wet warm weather are cured by this remedy. Kali Carb corrects the constitution of the patient and his susceptibility to cold disappears. Complaints of Kali Carb patients are aggravated between 3 to 5 A.M., the coolest period during night. The neuralgias of a Kali Carb patient appear to spread over different part of the body when it is cold, and if the affected part is kept warm, the pain goes to some other place which is uncovered and cold.

SOME OF THE REMEDIES WHICH SHOW AGGRAVATION FROM WET COLD WEATHER

Calcarea Carbonica:

The Calcarea patient has a characteristic sense of coldness. He gets cold sensation in feet and legs as if he had cold and damp stockings on. He has internal and external sensation of coldness of various parts of the head as if a piece of ice was lying against it. He is chilly patient and wants plenty of clothing. The Calcarea Carb patient takes cold very easily. Complaints of Calcarea Carb patients are aggravated in the rainy season but they keep well in cold climate provided it is dry. Calcarea Carb is suitable for symptoms that result from working in cold water. The patients sweat profusely. It is especially indicated, if other symptoms agreeing, in sweats of male organs, nape of neck, chest, axilla, hands, knees, feet, etc.

Dulcamara:

Complaints are caused or aggravated by change of weather from dry and warm to wet and cold. An individual who has been out in the heat of the sun resulting in profuse sweat, suddenly gets wet in rain resulting in suppression of sweats, may manifest a variety of symptoms which will respond to Dulcamara. Men working in ice-cream factories and cold storage houses go out into the sun and take some heat and then go back into their cold rooms and handle ice. These men are subject at times to diarrhoea and other catarrhal manifestations. Dulcamara will cure such patients. Complaints of Dulcamara are ameliorated in dry weather and by warmth in general. Dulcamara is for the manifestations from damp cold, while aconite is for the manifestations from dry cold.

Rhododendron:

The strongest characteristics of Rhododendron is in its modality, "aggravation in wet stormy weather". Rhododendron is particularly worse before a thunderstorm; after the storm breaks, the patient feels better. The aggravation before a thunderstorm does not at all depend on the dampness or coldness, but on the electrical condition of the atmosphere.

Rhus Toxicodendron:

The complaints of this remedy come from cold damp weather or from being exposed to cold damp air when perspiring. All his complaints are aggravated from wet and cold weather and are ameliorated from dry and warm air or weather. Symptoms of rheumatic character involving muscles of the leg, back and even spinal membranes caused by exposure to wet cold air, or by sleeping on damp ground or in bed with damp sheets or getting wet in a rain especially while perspiring, readily respond to this remedy.

Other notable wet weather remedies are *Natrum Sulphuricum* (humid asthma), *Nux Moschata* and *Calcarea Phosphorica* (wet cold, especially melting snow).

WEATHER SEQUENCE ASSOCIATED WITH WESTERN DISTURBANCE AND ITS ROLE IN SELECTION OF HOMOEOPATHIC REMEDIES

During winter, premonsoon and post-monsoon months western disturbances move across north India. On many occasions these disturbances appear on sea level charts as closed lows with frontal characteristics associated with them. Notwithstanding the fact that the delineation as well as the evolution of cold and warm fronts over India are to some extent subjective, these fronts can always be located as the boundary zones of two air masses. Ahead of the warm and cold fronts, moist and cold southeasterly, southerly and southwesterly winds generally prevail over north India. These disturbances at times give fairly widespread rain ahead of the warm and cold fronts. Complaints like croup, cold, coryza, influenza, asthma, bronchitis, rheumatism etc. which are caused and aggravated ahead of the cold and warm fronts may require some of the wet weather remedies mentioned earlier. Immediately after the passage of the cold front dry and cold winds invade north India. Complaints caused and aggravated due to exposure to cold and dry winds immediately after the passage of the cold front may require some of the cold dry weather remedies depending upon the totality of symptoms.

CONCLUDING REMARKS

In homoeopathy remedies are selected on the basis of the totality of symptoms. Though outwardly some of the symptoms appear queer, they are in reality manifestations of genetic characteristics of the individual. The response of the human body to weather changes is dependent on the constitution of the individual. Therefore it is no wonder that the susceptibility of an individual to weather changes plays a significant role in the selection of homoeopathic remedies.

REFERENCES

ALLEN, H.C. (1970): Characteristics of some of the Leading Remedies. Roy Publishing House, Calcutta-14.

KENT, J.T. (1966): Lectures on Homoeopathic Materia Medica. Roy Publishing House, Calcutta-14.

KENT, J.T. (1974): Repertory of the Homoeopathic Materica Medica. M/s. Jain Publishing Co., New Delhi-55. (Indian Edition).

LANDSBERG, H.E. (1972): The Assessment of Human Bioclimate. WMO Technical Note No. 123.

NASH, E.B. (1962): Leaders in Homoeopathic Therapeutics. Roysingh & Company, Calcutta-14.

SARGENT, F., and TROMP, S.W. (1964): A Survey of Human Biometeorology. WMO Technical Note No. 65.

ENVIRONMENTAL FACTORS ON ADRENOCORTICAL FUNCTION

R. Chandramouli and S. Subramanyam*
(Department of Physiology,
Kempegowda Institute of Medical Sciences,
K.R. Road, Bangalore - 560 004)

Abstract: - Psychosocial stimuli are the most potent stimuli that affect the hypothalamo - pituitary-adrenal system. The investigation of the hypothalamopituitary-adrenocortical activity in acute schizophrenia has been undertaken in the present study.

INTRODUCTION

The secretion of cortisol from the adrenal cortex is under the Control of Corticotrophin (ACTH) which in turn is regulated by the secretion of corticotrophin releasing factor (CRF). The CRF cells are located within the hypothalamus which comes under the regulatory influence of the limbic system and the neocortex. The main stimuli to CRF and ACTH secretion are stress and those environmental cues that entrain the intrinsic circadian rhythm. Mason (1968) has pointed out that psychological influences are the most potent neural stimuli that affect the pituitary-adrenocortical system. Psychological stresses represent a diverse group of stimuli producing emotional disequilibrium. Such stimuli include fear, anxiety, anger etc. The investigation of the hypothalamo-pituitary-adrenocortical activity in psychoses has been studied by several workers (Mason, 1968; Sachar et al., 1963; Seligman, 1975). In the present study we have investigated the adrenocortical activity in acute schizophrenics.

MATERIALS AND METHODS

The cases were selected from the Institute of Mental Health, Madras. Acute schizophrenics who were passing through two different clinical sequences of psychosis ultimately leading to recovery, were chosen (Mason, 1968). In each clinical sequence ten cases were selected.

Various terms have been used to describe the different stages of schizophrenia. The acute phase (Psychotic turmoil) may pass through

* Voluntary Health Services, Adyar Madras-20.

the stage of organic psychosis into anaclitic depression. Alternatively the psychotic turmoil may pass to stage A (parasitic phase) and then to stage B (compliance phase) leading eventually to recovery.

Plasma Cortisol was estimated by the method of Mattingly (1962). The estimations were done both in the morning (8AM) and in the evening (8PM) samples. In the urine 17OHCS (17-hydroxycorticosteroids), and 17KS (17-oxosteroids) by the method of Glenn and Nelson, (1953) 17KGS (17-oxogenic corticosteroids) by the method of Appleby (1955) were estimated.

RESULTS

Plasma Cortisol in Acute schizophrenics:

Plasma Cortisol in 8PM samples showed significantly higher level ($P<0.01$) than the morning samples (8AM). (Fig. 1). It is expected that in subjects with a normal sleep pattern, 8AM samples would

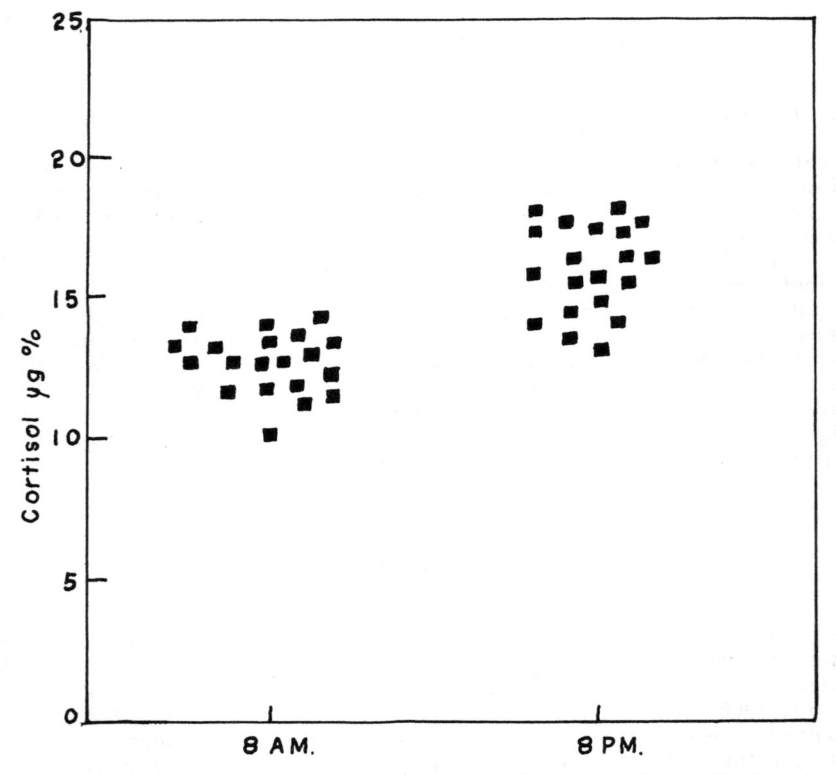

Figure 1: *Plasma cortisol in 8AM and 8PM samples. Each square indicates data of one patient.*

TABLE 1: *Urinary excretion of 17KS, 17OHCS, 17KGS in a clinical sequence of recovery from acute psychotic turmoil.*

	17KS mg/day	17OHCS mg/day	17KGS mg/day
Psychotic turmoil	21.41 ± 2.3	15.12 ± 2.2	22.75 ± 1.8
Organised Psychosis	10.67 ± 2.6*	8.34 ± 1.8*	14.22 ± 1.7*
Anaclitic Depression	15.14 ± 2.2	9.41 ± 2.3	17.52 ± 2.3
Recovery	10.28 ± 1.7**	7.83 ± 1.8**	13.28 ± 2.7**

Values are mean ± S.E.M.
* $P < 0.01$ compared with psychotic turmoil
** $P < 0.01$ compared with psychotic turmoil

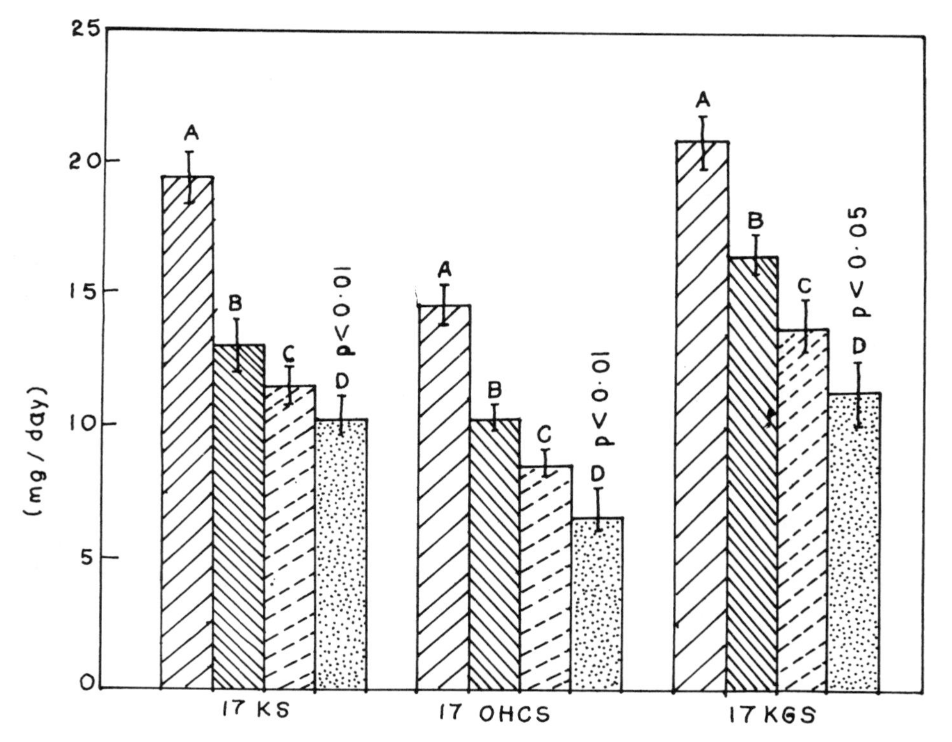

Figure 2: *Urinary excretion of 17KS and 17OHCS and 17KGS in a clinical sequence of recovery from acute psychotic turmoil.*
A) Psychotic turmoil. B) Stage A (Parasitic phase)
C) Stage B (complaint phase) D) Recovery
$p < 0.01$, Compared with psychotic turmoil.
$p < 0.05$, Compared with psychotic turmoil.

indicate the peak of the daily circadian curve. But sleep was so disturbed in these patients that their circadian rhythm was also disrupted.

Urinary Corticosteroid levels in the different stages:

Table 1 and Fig. 2 show that the Urinary 17KS, 17OHCS, and 17KGS levels were higher in the stage of acute psychotic turmoil. As they entered the phase of recovery the reduction in the urinary corticosteroids became obvious. From Table 1, it can be seen that the mean level of excretion of 17-OHCS in psychotic turmoil was roughly 200% higher than after recovery.

DISCUSSION

The results of the present investigation in acute schizophrenics bear out the speculation of Mann and Semrad (1969) and confirm the findings of the earlier study by Sachar et al. (1963). The state of acute psychotic turmoil in schizophrenics is associated with hypersecretion of the adrenal cortex. Studies on normal subjects undergoing prolonged psychological stress suggest that persistently increased adrenocortical activity occur only in a few, since most of them adjust to the situation (Mason, 1968). In acute schizophrenia this adjustment does not occur especially in the stage of psychotic turmoil, suggesting that there may be some primary hypothalamic dysfunction. Marked decrease in corticosteroids towards normal level in the phase of organised psychosis indicates the return of endocrine functions to normal. There is an increase in the corticosteroid excretion in the anaclitic depression phase compared with the phase of organised psychosis. The reason for the elevation could be interpreted in terms of the difference in the affective behaviour.

In the alternative pathway to recovery from psychotic turmoil, there were diminished urinary corticosteroid levels during the parasitic and compliance phases. The reason may be attributed to the differences in the degree or intensity of anxiety.

The disruption in the circadian rhythm of plasma cortisol in acute schizophrenics may be due to emotional arousal and also to disturbed sleep. The primary mechanism responsible for the increased adrenocortical activity in acute schizophrenia is likely to be the influence of limbic system and of structures involved in emotional states on the hypothalamo-neuro-endocrine centers.

REFERENCES

MASON, J. (1968): J. Psychosom. Med. 30: 563.

SACHAR, E. et al. (1963): J. Psychosom. Med. 25: 510.

SELIGMAN, M.E.P. (1975): In: "Helplessness in Depression, Development and Death". W.H. Freeman and Company, San Francisco.

MATTINGLY, D. (1962): J. Clin. Path. 15: 374.

GLENN, E.M. and NELSON, D.H. (1953): J. Clin. Endocrinol. 13: 911.

APPLEBY, T.J.I. et al. (1955): Biochem. J. 60: 453.

MANN, J. and SEMRAD, E. (1969): In: "Conversion as process and conversion as symptoms in psychosis". International Univ. press, New York.

SEASONAL VARIATION OF BMR IN HEALTHY NORMAL AND HYPNOTISED SUBJECTS*

H. Jana and R.B. Prajapati**
(Department of Physiology and Biochemistry, Smt. N.H.L. Municipal Medical College, Ahmedabad - 380 006)

Abstract: - BMR was determined in medical students of 18-23 years of age, during summer, monsoon and winter at Manipal, South Canara, Karnataka and at Ahmedabad (Gujarat). BMR in winter was significantly higher than during summer and monsoon. BMR in monsoon was significantly higher than during summer at Manipal, where the monsoon is very heavy and the temperatures during monsoon were not different from those during winter. On the other hand BMR in monsoon was significantly lower than during summer at Ahmedabad, where the monsoon is scanty. Some of the subjects were trained to undergo a moderate depth of hypnosis by preliminary talk, postural sway and suggestion of relaxation. BMR determined in the state of hypnosis in the three different seasons exhibited the same trend as in the waking subjects. Pulse rate, blood pressure, respiration rate, tidal volume, pulmonary ventilation and oral temperature of subjects did not differ in different seasons.

INTRODUCTION

Studies on energy metabolism in human beings is fairly common in different parts of the Indian subcontinent with the idea of having Indian standards of BMR at rest and during various activities, as a part of nutritional studies and as diagnostic aid. In almost all studies the indirect calorimetric method involving determination of O_2 consumption has been employed. In certain studies open circuit method using Douglas Bag and in Knipping's Recording Spirometer (closed circuit method), CO_2 elimination as well as O_2 consumption are determined to arrive at more accurate metabolic rate. Research workers in this field are quite aware of the sources of errors in the observations of BMR in human subjects. Unless one is very particular in training the subject with the type of mouth breathing, in taking

---

* The work was done at Kasturba Medical College, Manipal, South Canara, Karnataka during 1961-63 and at Smt. N.H.L. Municipal Medical College, Ahmedabad, during 1978-1979.

** Present Address: Department of Physiology, Anand Homeopathic Medical College and Hospital, Anand, Gujarat.

utmost precautions at every step of the procedure and rigid standards leading to accuracy are taken (Prajapati 1979) and the subject is taking part in this experiment preferably for a few consecutive days, very reliable valid metabolic rate cannot be arrived at.

In view of these possible errors, the reports on metabolic studies from various laboratories show quite a big range and are inconsistent. Considering the various climatic conditions, food habits of people of the different localities in India, and the adaptability of individuals to so many varying climatic conditions, the variation of BMR in different season and places in India having distinct seasonal variation is quite predictable. Variations in the consumption of food during different seasons are well known (Malhotra et al., 1976).

Malhotra (1960) reported that climate had no influence either on the resting or working metabolism. However, due to the additional clothing worn in winter, the total energy expenditure may be a little higher in winter. Davis and Johnson (1961) reported a highly significant change in shivering and a less significant change in heat production, and they suggested that man is seasonally acclimatized to cold and that this acquired acclimatization is least during summer months.

The paper presents some of the observations of BMR studies carried out during prominent seasons at Manipal in 1961-63 and at Ahmedabad in 1978-1979.

MATERIALS AND METHODS

At Kasturba Medical College, Manipal, 35 medical students (30 males and 5 females) of age-group 18-23 years, height 154.9-177.8 cm, body weight 33.75-73.8 kg and body surface area 1.25-1.9 m^2 volunteered for basal metabolic studies throughout the year. They were called for BMR determination during the month of April-May (Summer), July-August (Monsoon) and December-January (winter) in 1961-1963. Nineteen of them were trained hypnotic subjects; their metabolic rates were also determined during hypnotic trance state as well during the three seasons mentioned (Jana, 1964).

Knipping's Recording Spirometer was used for the determination of O_2 consumption and CO_2 elimination. The expired air was bubbled through strong (47%) potassium hydroxide solution in a wash bottle where carbon dioxide was absorbed. BMR was calculated based on the O_2 consumption taking RQ into consideration (Sterling, 1956).

The method of Hypnotic induction consisted of preliminary talk about pros and cons of hypnosis, some susceptibility tests like postural sway and hand-clasping and suggestion of relaxation and sleep (Jana, 1971).

At Smt. N.H.L. Municipal Medical College, Ahmedabad, 15 medical students (11 males and 4 females) of age-group 18-26 years, height 152-178 cm, body weight 38-60 kg and body surface area 1.3-1.76 m^2 were the subjects for the BMR estimation in all three prominent seasons of Ahmedabad - summer, monsoon and winter. Closed circuit Benedict-Roth apparatus with sodalime as the CO_2 absorbent was used for metabolic rate determination. Subjects maintained in basal condition were given rest for 30 min in bed before the spirogram was taken. Subjects were not allowed to come to lab by long walk or by cycling. Accuracy at the 5% level was strictly observed between the two observations in a day and between the averages of observations on consecutive days. R.Q. of 0.82, representative of the mixed diet, was assumed as carbon dioxide was not determined in the experiment.

Subjects were taken for BMR records in the following order:

4 subjects: Winter - Summer - Monsoon
4 subjects: Summer - Monsoon - Winter
7 subjects: Monsoon - Winter - Summer.

In this experiment subjects acted as their own control. The difference in paired values was used to determine statistical significance by fisher test.

RESULTS

TABLE 1: *Meteorological data of Manipal in 1978-1979.*

	SUMMER	MONSOON	WINTER
Temperature (°C)	27-31.5	24-28	23-28
Humidity (%)	75-93	85-97	59-77

TABLE 2: *BMR of same subjects in different seasons at Manipal.*

n	BMR ($KCal/m^2/hr$)	BMR ($KCal/m^2/hr$)	Significance
11	33.31 ± 1.805 (S)	34.81 ± 1.56 (W)	S Vs W $p < 0.01$
10	32.93 ± 1.434 (S)	33.79 ± 1.279 (M)	M Vs S $p < 0.001$
12	33.69 ± 1.178 (M)	34.61 ± 1.496 (W)	M Vs W $p < 0.01$

BMR in winter was significantly higher in comparison to summer and monsoon, both at Manipal and Ahmedabad. The relationship of the BMR in monsoon to that in summer was reverse in the two places of study. At Manipal BMR during monsoon (33.79) was 2.6% higher than the BMR during summer (32.93); the difference reached significance ($p < 0.01$). At Ahmedabad summer BMR (35.71) was 2.17% higher than monsoon BMR (34.95) but the difference did not reach significance ($p < 0.1$).

In studies at Manipal, the results of those subjects who volunteered in one season only are not shown in the table, but their data helped to confirm the trend in BMR during the three seasons. Subjects under neutral hypnosis (no specific suggestion was given during the trance) showed similar trends in metabolic rates as was seen in the waking

TABLE 3: *Meteorological data of Ahmedabad in 1978 - 1979.*

	SUMMER	MONSOON	WINTER
Maximum Temp. (C°)	35.7-46.2	25.8-33.2	26.0-32.3
Minimum Temp. (C°)	19.2-30.0	23.3-26.1	8.6-17.0
Humidity %	31 -90	84 -100	26 -84

TABLE 4: *BMR ($KCal/m^2/hr$) of 15 subjects in different seasons at Ahmedabad.*

Season	Mean	Standard Deviation	Significance	
Summer	35.71	0.989	Winter Vs Summer	$p < 0.001$
Monsoon	34.95	0.928	Winter Vs Monsoon	$p < 0.001$
Winter	37.32	0.855	Summer Vs Monsoon	$p < 0.1$

subjects. Metabolic rates in hypnotic trance state was akin to that in waking in the basal state; it was not lowered as in sleep (Jana, 1975).

The authors observed that cardiorespiratory parameters of the subjects at Ahmedabad did not vary significantly with seasons (Table 5). But BMR in winter was 4.5% higher than the BMR in summer and 6.8% higher than the BMR in monsoon.

TABLE 5: *Cardiorespiratory data in 15 subjects at Ahmedabad.*

	Summer	Monsoon	Winter
Oral Temperature (°F)	97.77	97.13	97.25
Pulse rate/min.	67.41	65.39	66.52
Systolic B.P. (mm Hg)	105.53	103.61	104.83
Diastolic B.P. (mm Hg)	60.72	58.33*	61.83*
Respiration rate/min.	14.18	13.63	14.19
Tidal volume (C.C.)	647.58	619.31	625.31
Pulmonary Ventilation (l/min.)	8.24	7.86	8.33

* Diastolic B.P.: Winter Vs Monsoon $p < 0.02$
All others : N.S.

DISCUSSION

In the present studies strict adherence to the basal condition, keeping early morning as the time of the experiments, appraising the subjects of the nature of experiments, trial of mouth-breathing, observance of silence in the laboratory and allowing only 5% fluctuation in the readings of the two observations and continuation of the experiment on consecutive days, enabled more accurate determination of BMR. That led to the conclusive evidence of seasonal variation in metabolic rate in the two places. In other centres, attempts to note the variation in BMR (mostly one day experiment) with seasonal variations were not successful. In this study the authors could obtain steady BMR value of a subject by 3 - 5 days of experimentation.

In Manipal where the ambient temperature range during monsoon and winter were practically the same, with variation in humidity, BMR in winter was highest of the three BMR values in summer, monsoon and winter. Again BMR during monsoon was significantly higher than during summer. It is worth mentioning that Manipal has quite heavy rainfall. At Ahmedabad, on the other hand, though winter BMR topped the three with greater significance, there was no significant difference between monsoon and summer BMR values. Ahmedabad has scanty rainfall.

From the study of the meteorological data with BMR values of subjects in different seasons, it is evident that it is the change in ambient temperature which influences BMR and that humidity does not apparently influence BMR.

Increased O_2 consumption by the subjects in winter is not mediated by increase in rate of respiration, increase in depth of respiration or by increase in pulmonary ventilation, because rate of respiration, tidal volume and pulmonary ventilation do not significantly vary during different seasons (Table 5). Hence the increased oxygen consumption during winter may be met by increased diffusion capacity of lungs, which is one possible mechanism for increased oxygenation, as suggested by Das and Jana (1978) or by increase in cross-sectional area of lungs as it occurs during pregnancy, reported by Novy and Miles (1967).

It is interesting to note that during winter there is increased in diastolic blood pressure (though not significant), which indicates increased tone of muscles of blood vessels. This finding lends support to the observation by Mukammal, Mckay and Neumann that at a given activity level and clothing insulation, there is higher incidence of coronary diseases when the thermal sensation is to the cold side of neutral comfort condition.

The Ama (diving women of Korea and Japan) showed the highest metabolic rate (35% above normal) in winter while the non-diving control natives showed a constant BMR throughout the year. This increase in metabolic rate could not be attributed to diet and appeared to be causally related to the degree of cold stress (Folk. 1974).

CONCLUSION

Seasonal variation of BMR was noted in young healthy students of Manipal and of Ahmedabad. In winter the BMR values were significantly higher in comparison to those during summer and monsoon.

ACKNOWLEDGEMENTS

The authors are grateful to the Dean, Kasturba Medical College, Manipal and to the Dean, Smt. N.H.L. Municipal Medical College, Ahmedabad for offering facilities for the work. Thanks are also due to Sri S.R. Chakraborty, Sri T.K. Das and Sri M.D. Pujara for their help in the work. The authors are thankful to Sri Ashvin V. Soni and Sri Ashvin M. Pandya for their continued help in the preparation of the manuscript.

REFERENCES

BEST, C.H. and TAYLOR, N.B. (1969): The Physiological basis of medical practice. 8th Ed., 2nd reprinting, Scientific Book Agency, Calcutta, 1276-1282.

DAS, T.K. and JANA, H. (1978): Basal metabolic studies during uterine cycle, Proc. 65th Session of the Indian Science Congress Association: 154.

DAVIS, R.A. and JOHNSON, D.R. (1961): Seasonal acclimatization to cold in man. J. Appl. Physiol., 16(2): 231-234.

FOLK, E.G. Jr. (1974): Textbook of environmental physiology, 2nd Ed., Lea and Febiger, Philadelphia, 180.

JANA, H. (1964): Energy metabolism in hypnotic trance and sleep. J. Appl. Physiol. 20(2): 308-310.

JANA, H. (1971): Physiological aspects of the Hypnotic trance state. Doctoral Thesis, University of Calcutta, Calcutta, 18-21.

JANA, H. (1975): Experience with student volunteers and patients in hypnosis. In: Therapy in Psychosomatic Medicine. F. Antonelli (Ed.); Edizioni Luigi Porri S.p.A. Roma, 443.

MALHOTRA, M.S. (1960): Effect of environmental temperature on work and resting metabolism. J. Appl. Physiol., 15(5): 769-770.

MALHOTRA, M.S. et al. (1976): Food intake and energy expenditure of Indian troops in training. Br. J. Nutr., 35: 229-233.

PRAJAPATI, R.B. (1979): Seasonal variation of basal metabolic rate at Ahmedabad. Postgraduate Dissertation, Gujarat University, Ahmedabad-22.

NOVY, M.J. and MILES, J.E. (1967): Respiratory problems in pregnancies. Amer. J. Obstet. Gynec., 99(7): 1026.

MUKAMMAL E.I., McKAY, G.A. and NEUMANN, H.H. (1984): A note on cardiovascular diseases and physical aspects of the environment. Int. J. Biometeor., 28(1): 17-28.

STERLING, E.H. (1956): Principles of human physiology. 12th Ed., H. Davson and M.G. Eggleton (eds.); Churchill, London, 810-811.

DYNAMIC LUNG FUNCTION STUDY OF CHRONIC BRONCHITIS PATIENTS ABOVE 40 YEARS

S. Chatterjee, B.P. Chatterjee, Dipali Saha and S. Das
(Work and Sports Physiology Laboratory
Department of Physiology Calcutta University College of Science
92, Acharya Prafulla Chandra Road
Calcutta 700 009, India)

Abstract: - Dynamic lung function tests were performed on 143 subjects. Amongst these subjects, 35 were normal nonsmokers, 43 were normal smokers and the rest were chronic bronchitis patients. Dynamic lung function studies included forced vital capacity (FVC), forced expiratory volume in 1 s (FEV_1), forced expiratory volume in 1 s as the percentage of forced vital capacity ($FEV_{1\%}$), forced expiratory time (FET), forced expiratory time of last 0.5 l ($FET_{0.5\ L}$), maximum voluntary ventilation (MVV), maximal mid-expiratory flow rate ($FEF_{25-75\%}$), maximal expiratory flow rate ($FEF_{200-1200\ ml}$), end-expiratory flow rate ($FEF_{75-85\%}$), peak expiratory flow rate (PEFR) and flow of last 0.5 l. Except PEFR, all the dynamic lung function tests were measured by Standard spirometric technique and PEFR was studied by Wright peak flow meter. All these parameters were measured in the standing posture and expressed in BTPS. Values of all the dynamic lung function tests in chronic bronchitis patients have deteriorated considerably in comparison with normal nonsmokers and smokers. The decrease in FEV_1, $FEV_{1\%}$, $FEF_{25-75\%}$, $FEF_{200-1200}$ ml, MVV and PEFR were highly significant and the increase in FET was highly significant ($P < 0.001$). It was also noted that the different values of dynamic lung functions tests were highly correlated amongst each other. It may be concluded that in chronic bronchitis patients, the dynamic lung functions are greatly impaired and the deterioration is directly related to the extent of smoking rate and increase in age.

INTRODUCTION

Chronic bronchitis is defined as a disorder characterized by hypersecretion of bronchial mucus, usually accompanied by chronic cough or recurrent productive cough for a minimum of three months in a year for at least two successive years, in patients in whom other causes for these symptoms have been excluded (American Thoracic Society, 1962). It is one of the chronic obstructive lung diseases. Chronic obstructive lung disease is associated with abnormalities in pulmonary ventilation, perfusion, diffusion and with nonuniformity of ventilation and perfusion through the lungs (Burrows et al., 1965). These disturbances in lung function are reflected in the result of a variety of pulmonary function tests.

Most researchers investigated the relationship of chronic bronchitis with smokers and established the fact that chronic bronchitis has a direct relationship with the extent of cigarette smoking (Geisler, 1969; Lambert and Reid, 1970; Durda et al., 1970; Armasu et al., 1971). In India, Radha et al. (1977) reported the results of a survey of prevalence of chronic bronchitis in an urban locality of New Delhi. Brinkman et al. (1963) investigated FEV_1, $FEV_{1\%}$ and MMEF of chronic bronchitis patients. Burrows et al. (1965) studied the intercorrelation of pulmonary function data in 175 cases of chronic obstructive lung disease. But the different aspects of spirometric study and the intercorrelation of pulmonary function data have not yet been studied properly in chronic bronchitis patients in our country and abroad.

So the objectives of our study are (1) to evaluate the dynamic lung functions of chronic bronchitis patients above 40 years of age when different types of chronic diseases of chest become very prominent; (2) to investigate the extent of deterioration in the said patients with respect to normal subjects (both non-smokers and smokers); (3) to find out the correlation amongst the different parameters of dynamic lung function tests and (4) to present regression formulae for some closely related variables.

MATERIALS AND METHODS

143 subjects were investigated for their dynamic lung function tests. Amongst these subjects 35 were normal nonsmokers, 43 were normal smokers and 65 were chronic bronchitis patients. The normal subjects were lecturers, staffs and technicians of Calcutta University of different departments and their healthy relatives and friends. The patients were provided by the Garden Reach Railway Hospital. The subjects were judged to be healthy on the following criteria.

(1) No history, current or past, of any cardiopulmonary disorders;
(2) No evidence of cardiopulmonary disease from physical examination, chest roentgenogram and electrocardiogram;
(3) No obvious signs of weakness or debility which significantly limits activity;
(4) Capable of adequate co-operation during the tests.

These were similar to the criteria used by Leiner et al., (1963) and Kory et al., (1961).

The normal 'nonsmoker' group consisted of subjects who denied any cigarette consumption during their life time or who had only smoked occasionally or smoked less than 5 cigarettes per day. The normal 'smoker' group consisted of subjects who had a continuous history of smoking and smoked a minimum of 5 cigarettes daily for minimum 5 years. Most subjects were light (15 cigarettes/day) and moderate smokers (15 to 20 cigarettes/day) (Densen et al., 1969).

The chronic bronchitis patients were diagnosed on the basis of case history, clinical, physiological and radiological findings. With the cooperation of one of our physician co-workers the above mentioned examinations were performed at the Garden Reach Railway Hospital. After the investigations in this regard were over, the patients were sent to our laboratory for the evaluation of dynamic pulmonary function tests in various aspects. The chronic bronchitis patients were not separated in to nonsmoker and smoker groups as there was no significant change in the parameters studied in these two groups.

Dynamic lung function studies included forced vital capacity (FVC), forced expiratory volume in 1 s (FEV_1), forced expiratory volume in 1 s as the percentage of forced vital capacity ($FEV_{1\%}$), forced expiratory time (FET), forced expiratory time of last 0.5 l (FET 0.5 l), maximum voluntary ventilation (MVV), maximal expiratory flow rate ($FEF_{200-1200}$), maximal mild-expiratory flow rate ($FEF_{25-75\%}$), end-expiratory flow rate ($FEF_{75-85\%}$), peak expiratory flow rate (PEFR) and flow of last 0.5 l.

FVC, FEV_1, $FEV_{1\%}$, FET, $FET_{0.5\ l}$, MVV, $FEF_{25-75\%}$, $FEF_{75-85\%}$, $FEF_{200-1200\ ml}$ and flow of last 500 ml were measured by standard spirometric technique and PEFR was determined with the help of Wright Peak flow meter. All the above mentioned parameters were expressed in BTPS.

RESULTS

65 chronic bronchitis patients and 78 normal subjects were studied for their dynamic lung function tests. Among the 78 normal subjects, 35 were nonsmokers and the rest were smokers. The physical characteristics of bronchitis and normal subjects are shown in Table 1. The difference of age and height amongst the three groups was not significant but the weight of bronchitis patients was significantly lower in comparison with that of normal nonsmokers ($P < .001$) and smokers ($P < .01$).

Table 2 indicates the mean values and standard deviations of FVC, FEV_1, $FEV_{1\%}$, FET, $FET_{0.5L}$ of normal nonsmokers, smokers and chronic bronchitis patients.

FVC, FEV_1 and $FEV_{1\%}$ were decreased significantly in patients as compared to normal nonsmoker subjects ($P < .02$, $P < .001$ and $P < .001$ resp.) and these parameters except FVC were also deteriorated significantly in patients in comparison with those of normal smoker subjects. FET and $FET_{0.5L}$ increased significantly in patients when compared with those of both normal nonsmoker and smoker subjects.

Table 3 shows the mean values and standard deviation of MVV, $FEF_{25-75\%}$, $FEF_{200-1200\ ml}$, PEFR, $FEF_{75-85\%}$, flow of last 0.5 l of normal nonsmoker, smoker and chronic bronchitis patients. The decrease for all these parameters were highly significant when compared with those of normal nonsmoker and smoker subjects.

Table 4 indicates the correlation coefficients amongst the dynamic lung functions of chronic bronchitis patients. Fig. 1 and 2 show the relationship of FEV_1 with $FEF_{25-75\%}$, MVV and PEFR. FEV_1 correlated significantly with $FEV_{1\%}$, MVV, $FEF_{25-75\%}$, $FEF_{200-1200\ ml}$, $FEF_{75-85\%}$ and PEFR ($r = 0.70$, 0.69, 0.75, 0.79, 0.60 and 0.77 resp.). There were also significant correlations amongst the other forced expiratory

TABLE 1: *Mean values and standard deviations for physical characteristics of Normal and Bronchitis subjects above forty years of age.*

Parameters	Normal Nonsmoker n = 35	Normal Smoker n = 43	Bronchitis n = 65	Significance of differences in mean values between Bronchitis and Normal subjects	
				Bronchitis vs. Normal nonsmoker	Bronchitis vs. Normal smoker
Age (years)	48 ± 5.59	49 ± 6.19	50 ± 5.90	NS	NS
Height (cm)	163.6± 7.19	162.6± 7.60	164.5± 6.44	NS	NS
Weight (kg)	57 ±12.28	53.6±10.25	49 ± 8.13	$P < .001$	$P < .01$

TABLE 2: *Mean values and standard deviations for dynamic lung function tests of Normal and Bronchitis subjects above forty years of age.*

Parameters	Normal Nonsmoker n = 35	Normal Smoker n = 43	Bronchitis n = 65	Significance of differences in mean values between Bronchitis and Normal subjects	
				Bronchitis vs. Normal nonsmoker	Bronchitis vs. Normal smoker
FVC (l)	3.347±0.62	3.145±0.51	3.028± 0.55	$P < .02$	NS
FEV_1 (l)	2.689±0.50	2.369±0.42	1.918± 0.52	$P < .001$	$P < .001$
$FEV_{1\%}$	78 % ±6.20	76 % ±6.49	63 % ±10.94	$P < .001$	$P < .001$
FET (sec)	5.18±1.41	5.96±1.73	7.38± 2.44	$P < .001$	$P < .001$
FET 0.5 l (sec)	3.92±1.28	4.51±1.57	5.40± 1.99	$P < .001$	$P < .02$

TABLE 3: *Mean values and standard deviations for dynamic lung function tests of Normal and Bronchitis subjects above forty years of age.*

Parameters	Normal Nonsmoker n = 35	Normal Smoker n = 43	Bronchitis n = 65	Significance of differences in mean values between Bronchitis and Normal subjects	
				Bronchitis vs. Normal nonsmoker	Bronchitis vs. Normal smoker
$FEF_{25-75\%}$ (l/min)	202 ± 55.35	140 ± 55.65	89 ± 51.88	$P < .001$	$P < .001$
$FEF_{200-1200\ ml}$ (l/min)	286 ±101.59	253 ±100.57	162 ± 71.79	$P < .001$	$P < .001$
$FEF_{75-85\%}$ (l/min)	42.29± 10.44	31.94± 14.21	24.99± 13.40	$P < .001$	$P < .02$
MVV (l/min)	105 ± 20.45	88 ± 19.23	73 ± 23.67	$P < .001$	$P < .001$
PEFR (l/min)	519 ± 75.96	488 ± 74.97	404 ±117.20	$P < .001$	$P < .001$
Flow of last 0.5 l (l/min)	9.423± 4.32	8.347± 3.76	6.693± 2.34	$P < .001$	$P < .02$

TABLE 4: *Coefficients of correlation among dynamic lung function measurements in Bronchitis subjects.*

	FEV_1(l)	$FEV_{1\%}$	$FEF_{25-75\%}$ (l/min)	$FEF_{200-1200}$ (l/min)	$FEF_{75-85\%}$ (l/min)	MVV (l/min)	PEFR (l/min)
FEV_1(l)		0.70	0.75	0.79	0.60	0.69	0.77
$FEV_{1\%}$			0.68	0.72	.30	0.45	0.68
$FEF_{25-75\%}$ (l/min)				0.68	0.61	0.66	0.71
$FEF_{200-1200\ ml}$ (l/min)					0.41	0.57	0.62
$FEF_{75-85\%}$ (l/min)						0.61	0.63
MVV (l/min)							0.79
PEFR (l/min)							

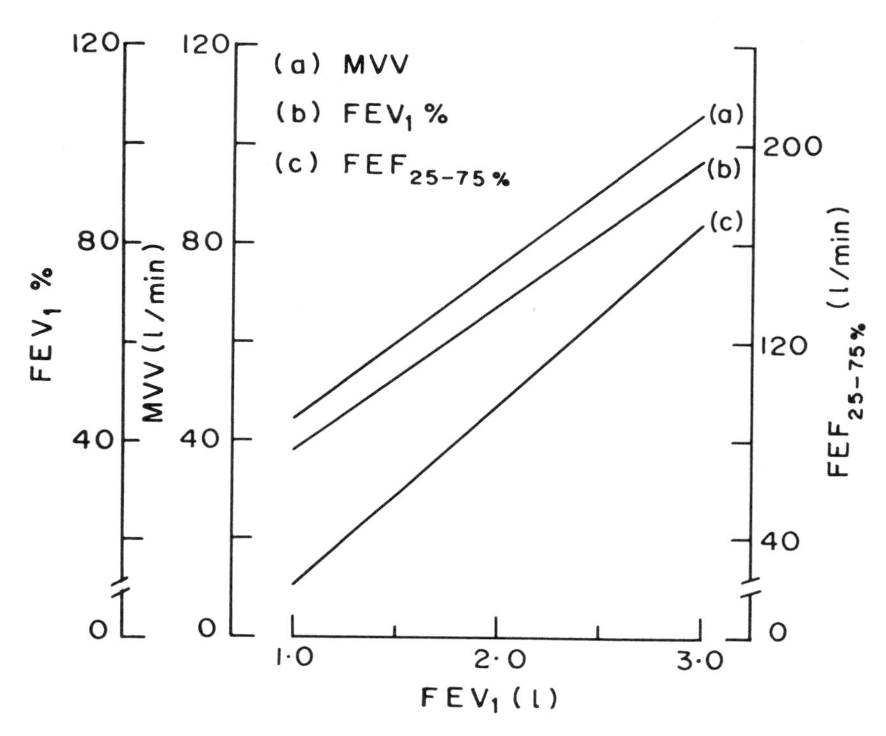

Figure 1: *Relationship of FEV_1 to MVV, $FEV_1\%$ and $FEF_{25-75\%}$ in Bronchitis patients.*

flow rates. On the basis of the close correlations, some regression formulae are given below:

Variables	Formulae	r
MVV (l/min) vs FEV_1(l)	MVV = 31.408 FEV_1 + 12.76	0.69
$FEV_{1\%}$ vs FEV_1 (l)	$FEV_{1\%}$ = 14.727 FEV_1 + 34.374	0.70
$FEF_{25-75\%}$(l/min) vs FEV_1(l)	$FEF_{25-75\%}$ = 74.827 FEV_1 - 54.52	0.75
$FEF_{200-1200\,ml}$(l/min) vs FEV_1(l)	$FEF_{200-1200\,ml}$ = 109.066 FEV_1 - 47.189	0.79
$FEF_{75-85\%}$(l/min) vs FEV_1(l)	$FEF_{75-85\%}$ = 15.462 FEV_1 - 4.666	0.60
PEFR (l/min) vs FEV_1(l)	PEFR = 173.546 FEV_1 + 71.139	0.77

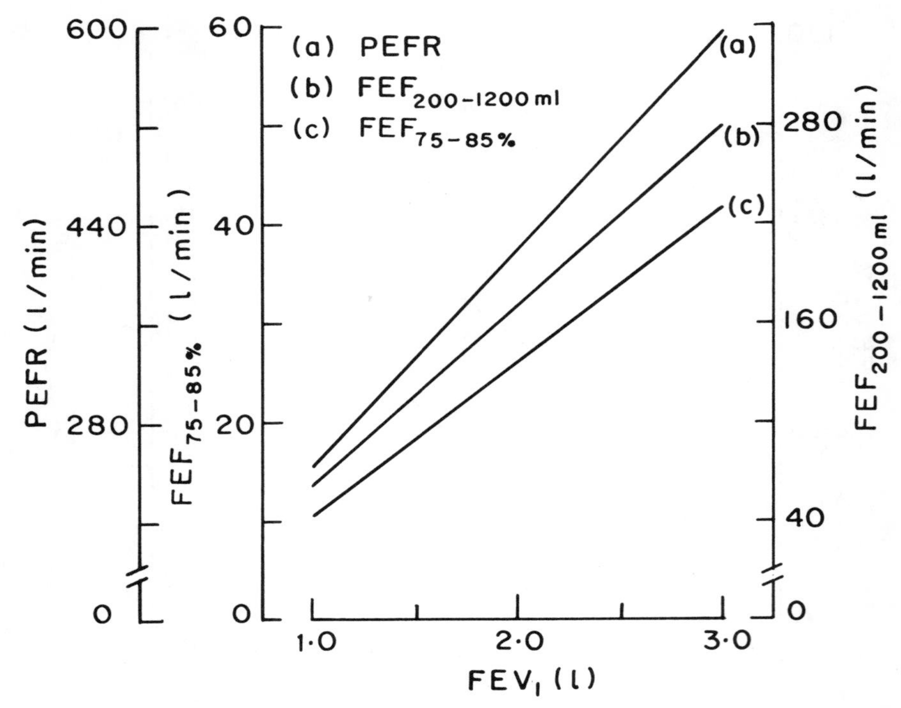

Figure 2: *Relationship of FEV_1 to $FEF_{200-1200\ ml}$, $FEF_{75-85\%}$ and PEFR in Bronchitis patients.*

DISCUSSION

Selected dynamic lung function tests were performed on 35 normal nonsmokers, 43 normal smokers and 65 chronic bronchitis patients. Amongst chronic bronchitis patients, 53 were smokers and only 12 were nonsmokers. Fletcher (1969) had presented a broad review of the evidence supporting a causal relationship between cigarette smoking and chronic respiratory disease, particularly emphysema and bronchitis. Geisler (1969) observed that most important exogenous factor in chronic bronchitis was undoubtedly cigarette smoking and this worker stated that severe bronchitis not preceded by a history of smoking was as rare as the finding of normal lung function values in heavy smokers.

We observed that FEV_1, $FEV_{1\%}$ and $FEF_{25-75\%}$ were adversely affected in chronic bronchitis patients. Brinkman et al. (1963) observed similar findings. Comroe et al. (1977) gave data on pulmonary function tests of a 39-year old bronchitis patient. The patient had been suffering from bronchitis as a result of inhalation of irritant fumes. They observed the values of MVV, $FEF_{200-1200\ ml}$, and $FEV_{1\%}$ as 90 l/min, 170 l min and 66% but we obtained these values as 73 l/min, 162 l/min and 63% respectively. The lower values obtained by us might be due to the higher age group of our subjects. Wysocki (1972) also noted that ventilatory function was impaired in men and women with manifestations of chronic bronchitis compared to healthy men and women.

We obtained close correlation amongst FEV_1, $FEF_{25-75\%}$ and MVV. Burrows et al. (1965) also noted similar observations in their study of 175 COPD patients. This was expected in as much as all these three measurements reflected forced ventilatory flow rates. In the case of FEV versus MVV, both of which measured ventilatory flow rates, the correlation was so close that many European investigators had reported FEV in terms of an "indirect maximum breathing capacity" [Burrows et al., 1965]. Our formula relating FEV_1 to MVV ($MVV = 31.408\ FEV_1 + 12.76$) was not similar to those utilized by Burrows et al. (1965) and Gandivea and Hugh-Jones (1957). [$MVV = 36.7\ FEV_1 + 1,5$ (Burrows et al., 1965), $MVV = 35$ or $37.5\ FEV_1$ (Gandivea and Hugh-Jones, 1957]. The difference in formulae might be due to the fact that they had considered the COPD patients but we had taken into consideration the chronic bronchitis patients only.

Thurlbeck (1975) observed that goblet cell metaplasia is an obvious feature of patients with chronic bronchitis and emphysema and it may be responsible for producing obstruction in peripheral airways of these subjects. Cullen (1972) investigated the cause and prevalence of chronic bronchitis in Busselon, a town relatively free from air pollution and industrial dusts. He reported that various factors had a significant relationship to this disorder, including smoking, breathlessness, recent chest illness, productive cough, exposure to dust occupations and gold mining. According to Peter Howard (1967) the chronic bronchitis is associated with some pathological process which is responsible for a steady erosion of ventilatory capacity. He also pointed out that the significance of bronchial infection is difficult to assess. The acute exacerbations so characteristic of the disease have been attributed to bronchial infection on clinical grounds and by the isolation of pathogenic organisms such as *Streptococcus pneumoniae* and *Hamophilus influenzae* from the sputum. Pathologists looking at the terminal phases of the disease have suggested that bronchial infection plays an important part in the development of destructive lung changes.

In conclusion we may say that the dynamic lung functions in chronic bronchitis patients would definitely be deteriorated. And we may also infer that higher the smoking rate and higher the age, the higher is the prevalence of chronic bronchitis.

ACKNOWLEDGEMENTS

Authors are grateful to UGC, New Delhi for financial assistance.

REFERENCES

ARMASU, C., BUMBACESCU, N., ARBORE, G., PIROZYNSKI, M., and BRINZEL, M. (1971): Date w privire la prevalenta bronsitei cronice. Rev. Med. Chir., 75: 861-869.

BRINKMAN, G.L., and COATES, E.O., Jr. (1963): The effect of bronchitis, smoking and occupation in ventilation. Amer. Rev. Resp. Dis., 87: 684-693.

BURROWS, B., STRAUSS, R.H. and NIDEN, A.H. (1965): Chronic obstructive lung disease. III. Interrelationships of pulmonary function data. Amer. Rev. Resp. Dis., 91: 861-868.

CHRONIC OBSTRUCTIVE LUNG DISEASE. A statement of the Committee on Therapy (1965): Amer. Rev. Resp. Dis., 92: 513-518.

COMROE, J.H., Jr., FORSTER, R.E., DUBOIS, A.B., BRISCOE, W.A. and CARLSEN, E. (1977): The lung. Clinical physiology and pulmonary function tests. Chicago: Year Book Medical Publishers, Inc.

CULLEN, K.J. (1972): Chronic bronchitis and the Australian environment. Med. J. Aust., 1: 249-253.

DENSEN, P.M., JONES, E.W., BASS, H.E., BREUER, J., and RECD, E. (1969): A survey of respiratory disease among New York City Postal and Transit Workers. 2. Ventilatory function test results. Environ. Res., 2: 277-296.

DURDA, M., SZAFRANSKI, W., SMIECHNAJDA, B., and KOSCIOLKO, L. (1970): Prezewlekle nieswoiste choroby pluc u pracownikow zakladow elektromaszynoruych "EDA" w Poniatowej. Gruzlica, 38: 623-634.

FLETCHER, C.M. (1969): Cigarettes and respiratory disease. Inhalation Ther., 14: 45-50.

GANDIVEA, B., and HUGH-JONES, P. (1957): Terminology for measurement of the vital capacity. Thorax, 12: 290.

GEISLER, L. (1969): Klinik der chronischen Bronchitis. Tägliche Praxis, 10: 377-389.

HOWARD, P. (1967): Evolution of the ventilatory capacity in chronic bronchitis. Brit. Med. J., 3: 392-395.

KORY, R.C., CALLAHAN, R., BOREN, H.G. and SYNER, J.C. (1961): The Veterns Administration - Army Co-operative Study of Pulmonary Function. Amer. J. Med., 30: 243-258.

LAMBERT, P.M. and REID, D.D. (1970): Smoking, air pollution and bronchitis in Britain. Lancet, 1: 853-857.

LEINER, G.C., ABBRAMOWITZ, S., SMALL, M.J., STENBY, V.B. and LEWIS, W.A. (1963): Expiratory peak flow rate. Amer. Rev. Resp. Dis., 88: 644-651.

RADHA, T.G., GUPTA, C.K., SINGH, A., and MATHUR, N. (1977): Chronic bronchitis in an urban locality of New Delhi, an epidemiological survey. Ind. Jour. Med. Res., 66: 273-285.

THURLBECK, W.M., MALAKA, D., and MURPHY, K. (1975): Globlel cells in the peripheral airways in chronic bronchitis. Amer. Rev. Resp. Dis., 112: 65-69.

WYSOCKI, M. (1972): Testy spirometryezne. TQTi FEV-wlasciwosci i przydatnose w roz poznawane 'v przewleklege zapaleni oskrzeli. Gruzlica, 40: 35-44.

SUBJECT INDEX

AUTHOR INDEX